Settings and Stray Paths

Settings and Stray Paths

Writings on Landscapes and Gardens

TO KAREN + JOHN.
FELLOW LANDSCAPE TRAVELERS.
WHO HELPED MAKE IT HAPPEN.
WITH BEST WISHES.

Marc

$$\frac{16}{VIII}{05}$$

BERKELEY

Marc Treib

LONDON AND NEW YORK

First published 2005 by Routledge

2 Park Square, Milton Park, Abingdon, Oxon OX14 4RN

Simultaneously published in the USA and Canada
by Routledge, 270 Madison Avenue, New York, NY 10016

Routledge is an imprint of the Taylor & Francis Group

Designed by Marc Treib

Typeset in Century Old Style and Bell Gothic Light

Printed and bound in Great Britain by
the Alden Press, Oxford

Photographs by the author, except as noted

British Library Cataloguing in Publication Data

A catalogue record for this book is available from the British Library

Library of Congress Cataloging in Publication Data

Treib, Marc.

  Settings and stray paths : writings on landscapes and gardens /
MarcTreib.

     p. cm.

Includes bibliographical references and index.

ISBN 0-415-70046-9 (hb : alk. paper)—ISBN 0-415-70047-7 (pb : alk.

paper)—ISBN 0-203-41282-6 (ebook) 1. Landscape architecture.

I.Title.

 SB472.4.T74 2005

 712--dc22

                                    2004021564

ISBN 0-415-70046-9 (hbk)

ISBN 0-415-70047-7 (pbk)

To Georges Descombes,
who practices thinking

Contents

Preface

This is not a book.

This is a collection of essays published in the form of a book.

They were written for different audiences and different publications
over a span of almost twenty-five years and, as such, they reflect an
evolution rather than a steady state. There are areas of overlap in these
essays, a considerable overlap in ideas, examples, and sources—
especially in a cluster of the early writings. Perhaps I overused the
Woodland Cemetery outside Stockholm, Michael Heizer's *Double
Negative* and other earthworks, and the gardens of Versailles and
Ryôanji—but at the time these were the best representatives I knew.
In some cases they remain so. Since the various writings were never
intended to be collected in one place, these repetitions are somewhat
unavoidable. Areas of gross repetition have been removed, but I beg
the reader's indulgence for the redundancies that remain.

Each of these essays is essentially an exploration of a thought or series
of thoughts. Most began with observations on form and space—
occasionally behavior—rather than with any abstract or theoretical
idea. It would seem that my normal path extends from the specific to
the general, or from the concrete to the abstract. The range of ideas,
perhaps, is not that broad, and certain topics recur in a number of the
essays. In the early writings, for example, the idea of order figures
prominently. Just why, I am not certain, but I suspect it stems from a
background in architecture—which is essentially composed and constructed—and an attempt to rebuff the almost universally accepted
Californian notion that raw nature is best. If I can dimly recall those
times some twenty-five years ago, I was trying to argue how the shaped
landscape—the one that is obviously designed and fabricated—while
perhaps less comforting as an escape, was actually more interesting as
a locus for thought and interpretation. And so, from the topic of order
the path traversed the thorny issues of the formal and informal landscapes
—and their intersection in Japanese garden design. From these explorations of syntax I hesitantly entered the more turbulent waters of
relativity and semantics. Whether my arguments are convincing or
not will be left to the reader, but in the mere provocation to thought
on his or her part I will assume some measure of success.

Reviewing one's writings from a long perspective can be both sobering and revealing. It would seem I have at times conflated—incorrectly perhaps—the experience of the natural, cultural, and designed landscapes, considering the effects to be somewhat similar whether aesthetic intention instigated the making of the landscape or not. I also can discern a stress on thinking about the landscape as well as its experiential dimension, that is, how the shape of the place relates to concepts and pleasure.

I would like to think that in these writings, and in others of course, my explication of ideas and critiques of realized landscapes spans the three arenas key to understanding landscapes: the social, the environmental, and the formal (as in space and materials rather than in degree of formality). Experience has shown that all too often authors advance only one side of the triad while excluding the others from their purview. The environmentalists might decry any use of a national forest and loathe topiary as an unnatural act. Those more socially oriented consider the addition of a bench a great humanistic act without considering the spatial or aesthetic consequences, while those more inclined to value composition and materials might not consider the human body and its comfort, or the human urge to gather in groups. While I tend to focus on the landscape as realized and perceived, the consideration of all three sets of factors inform my thinking. So do modernist values: appreciating an economy of means, and a complex experience achieved through simple conditions rather than vice versa. And behind it all lies the search for the poetic dimension that transcends the pragmatics of topography, climate, and use—something that graces landscape design with the air of art.

Introduction

I.

Selecting one's writings for a collection, like suffering through the introduction before a lecture, instigates thoughts of a "near death" experience: you can see your whole life (well, at least your professional life) flash before you. I probably never would have thought of making this modest anthology had it not been for the editor's invitation, and it has been done only with some pause, and with considerable advice from friends. It seemed too early in life for such an effort. Who would be interested in these articles, a number of which might be fairly termed "dated"? Which ones should be included? All of them seemed equally relevant or equally superannuated. I must admit that even at this point I am still not sure about the answer to either of these questions or to many others; but here is the collection nonetheless.

At the outset I need confess to seeing the world with a bias. Trained as an architect, and having worked over the years as a graphic designer as well as an educator, my interest has focused on form and space; places realized; physical environments created in response to social and environmental conditions; clever and/or beautiful and/or moving landscapes produced in relation to specific constraints. What is the difference between the intended use and the used intention, the intended perception and the perceived intention? In many instances it is not an abstract thought that provided the point of departure, but an observation: something has struck me as intelligent or absorbing or exquisite and I wonder about the process behind it. Just how did it get to be that way? What is its (hi)story? Who designed it and what was his or her story? In the reaction against formalism in architecture and design criticism we have somewhat lost the ability to describe, analyze, and critique form and space, that is, the actual stuff of the built environment. This is unfortunate. I can recall visits to many churches in Rome with an esteemed colleague who explained the various paintings solely in terms of iconography and political or religious intention. At one point I asked whether the fact that these were painted images (as opposed to written texts), and painted the way they were, and with what composition, and…Didn't those factors have any significance for an art historian? We may center our attention on the politics, social consequences, environmental performance, and a host of other factors —all of which are valid and necessary for understanding what we make and life within it. But we often never quite get to the actual thing that is the product of the process, that thing that appears, smells, sounds, or feels the way it does. This has been my principal arena.

The modernist (which I must admit to being—still) certainly appreciates antecedents from the past, and learns from its achievements. But there is an equally strong resolution against repeating the prior solutions. Instead, one attempts to understand the ideas behind those forms, to recast them in greater accord with contemporary conditions. At times complete invention is necessary and good; at other times, moderation offers a better path. There is no denying the historical importance of landscapes past, nor their lessons from which we can learn. But we can and must also add to landscape culture on our own societal terms. Hence a vision is needed that achieves more than merely addressing today's needs. In this I differ from a number of my colleagues who evaluate the importance of a landscape design primarily in terms of its immediate functional or social performance, in some ways considering the landscape architect's work as a type of social or environmental plumbing, denying the cultural implications of design. It is necessary to elevate pragmatic concerns to those poetic; not by avoiding social and environmental conditions but by achieving more than rote performance. In some ways this rehearses Le Corbusier's conclusion in *Towards a New Architecture*, in which he distinguishes between architecture and building:

> You employ stone, wood and concrete, and with these materials you build houses and palaces; that is construction. Ingenuity at work.
>
> But suddenly you touch my heart, you do me good, I am happy and I say: "This is beautiful." That is Architecture. Art enters in.[1]

Could we not make a similar distinction between landscaping or site remediation and landscape architecture? Can we not regard landscape architecture as a significant contribution to society's cultural as well as functional project?

Some still question the ability of landscape design to achieve the level of art. To a certain degree this is a personal concern; at other levels it is a matter of academic definition. In my mind, art is a quality rather than a class of object; it is that ability to strike me in some way beyond the norm, to move me, to make me consider that something is beautiful, or perhaps even edifying. Functional objects may achieve this quality of art; non-functional objects hold no such guarantee merely because they do not address use.

As a practice, however, the lines between design and art are more clearly drawn, at least in my mind. The designer starts with constraints external to him or herself; the artist begins with internal constraints; it is more about the self. Design as a total practice, of course, involves the internal considerations of the artist and art, and vice versa. (The realm of art commissions and public art occupy a never-never land spanning the two practices.) So landscape architecture must of necessity begin with those external parameters: the site, the climate, pragmatic issues such as drainage and temperature, social issues such as the number of occupants and their constituency, their nature, their projected activities—and of course the budget and resources at hand. But that is only the beginning of the work. The photographer Edward Weston once directed us to photograph a thing not for what it is, but for what else it is. That "what else" lies at the base of the best landscapes humans have produced. Perhaps the search for these places, and the search for the understanding of the conditions of their creation, lie at the root of my writing.

II.

The essays included in the book are only a fraction of those I have written, and one might question how the selections were made. For pragmatic reasons I did not include the text of lectures that have remained unpublished, nor chapters included in edited books. I restricted the body of work to those essays whose issues are still relevant to my thinking; to my favorite subjects, people, and places; and to those that have been used as references and in course readers. Several colleagues agreed to review the essays I had assembled as a sort of "rough cut." Considering their suggestions as well as inexplicable feelings, the final selections were made. While I have edited all of the essays to varying degrees, I have tried to maintain the original structure and ideas behind them. For economic reasons, in some cases the number of illustrations have been reduced. In other places, new and improved images have substituted for the originals.

III.

Then there is the problem of my standing today with an advanced state of knowledge (or so one would hope), with the essays frozen in time. Although I still support the body of the writings included here, there are certain points that no longer reflect my current thinking. Comments on these new positions, and in some instances new references, may follow at the end of the essay in the notes, set in italic type. Hopefully, this system will establish a distance, where necessary, between then and now, and at least partially defuse that embarrassing situation in which the ignorance of the author is only too apparent. In any event, despite all my hedging and creating of barriers against criticism, a collection of essays remains a columbarium of ideas, phrases, and illustrations that is as much a memorial as it is an assertive battle cry. I would hope that these writings might stimulate the interest in others to examine the history and practice of landscape architecture and to design landscapes that will provide us with both pleasure and provocation to thought.

Berkeley
July 2004

Note

1 Le Corbusier, *Towards a New Architecture*, Frederick Etchells, translator, 1931, reprint, New York: Dover Publications, 1986, p. 203.

[1]

Reduction, Elaboration, and Yûgen
The Garden of Saihô-ji

1989

With the years that have passed by
It has grown austere and holy
  On Mount Kageru:
The cedar tree, upright like a spear,
Already has a layer of moss at the root.[1]

To Zeami Motokiyu (1363–1443), a medieval Noh actor and theorist, the theater was a vehicle to achieve the revelation of *yûgen*. A term evading exact definition in Japanese as well as English, yûgen suggests a sense of mystery, depth; "elegance, calm, profundity, mixed with a feeling of mutability" [1-1].[2] The Noh actor, masked and controlled, conjured the feeling of yûgen through his understanding of nature and existence. Zeami used the metaphor of the landscape gardener to explain how the actor portrayed the essence of a phenomenon: the gardener sees and understands the river and recreates its spirit in the garden, though he does not replicate its exact form. Like the gardener, says Zeami, the Noh writer puts forth the "garden within his soul," in words and in movement.

Youth possesses exuberance and is quick to express its feelings. The passing of years, however, adds the depth requisite for achieving yûgen. This higher level is expressed in the Zen image of snow piled in a silver bowl. Higher still is the quality that surpasses perfect resolution: "Snow has covered thousands of mountains. Why is it that a solitary mountain towers unwhitened among them?"[3] This mountain—or actor or garden—achieves the level of profundity that rises above mere technical competence through wisdom and selflessness.

The garden at Saihô-ji, on the southwestern outskirts of Kyoto, evokes just this sense of yûgen. Bathed in shadow or dappled light, sheathed in moss, the full limits of its four and a half acres may be examined along two axes: the experiential and the religious. Saihô-ji, or the Temple of the Western Direction, may trace its origins to a sixth-century estate of the nobility, although no documented evidence supports this claim. By the twelfth century, the estate had been converted to a Buddhist temple of the Jôdo, or Pure Land sect.[4] In contrast to the more elaborate formal rituals of the earlier Esoteric and Exoteric sects of Japanese Buddhism, Jôdo offered a relatively simple doctrine for worship and achieving enlightenment. Central to its belief was the deity of the Amida Buddha, who in his Western celestial realm aided in the salvation of all sentient beings. Like the conception of paradise in many other lands, the heavenly sphere was conceived in contrast

[1-1]
SAIHÔ-JI. KYOTO,
JAPAN. CIRCA 1338.
MUSÔ SOSEKI.
The garden in winter
afternoon light.

to local climatic conditions and the drudgery and troubles of quotidian existence.

The gardens of the Pure Land sect were intended as re-creations or evocations of Amida's Western Paradise and hence are known as paradise gardens. Lushly planted rather than austere, using living material rather than rock and sand, they offered direct symbols in place of the suggestive abstraction of the later Zen sect. Saihô-ji utilized all these aspects and remains—in spite of later overlays—one of the most representative of the Japanese paradise gardens.

The precise state of the garden at the time of its creation about 1200 remains unknown, and its appearance today reflects to a considerable degree the later influence of Zen Buddhism. The pond—which is thought to have been sufficiently ample to support boating—and the basic layout, however, date from its origin [1-2]. In the villa's early years numerous structures dotted the hillside, but most of these met their demise in the fifteenth-century Ônin civil wars. Thus, the structures that greet the visitor today are of relatively, or very recent, construction and only two teahouses lay claim to a true history [1-3]. One of these, the Shonan-tei—thought to date to the Momoyama period (1568–1615)—is regarded as a national treasure.

Based on the Chinese Ch'an sect, Zen was imported into Japan in the twelfth century and soon gathered a considerable following, particularly among the samurai, or military class, which admired the new sect's acceptance of the toils of existence and the austerity and regulation its doctrine demanded. As Pure Land Buddhism had developed while acknowledging the earlier, more involved, orthodoxy, Zen itself reduced formal ritual and relied instead on the individual, with meditation as the primary means for encouraging enlightenment. While *satori,* or enlightenment, arrived in an instant, and was not achieved through a logical process of study alone, one could create settings conducive to meditation. The temple's principal garden, usually set adjacent to the south side of the Zen hall, supported contemplative practice, and in turn perhaps the attainment of the ultimate goal.

Musô Soseki (also known as Musô Kokoshi) played an important role in the creation of several gardens and in shaping the doctrine

[1-2]
SAIHÔ-JI.
The Golden Pond, islets, and convoluted shores.

[1-3]
SAIHÔ-JI. THE SHONAN-TEI (TEA HOUSE).

that reinforced the position of the garden in the Zen sect. A priest who lived from 1275–1351, Musô himself was an avid gardener who believed that the making and tending of the garden was in no way secondary to its contemplation. The garden condensed the features and the spirit of the landscape, and the landscape of the spirit.

> Still others see the mountain, the river, the earth, the grass, the tree, the tile, the pebble, as their own essential nature. They love, for the length of the morning, the mountain and the river. What appears in them to be no different from a worldly passion is at once the spirit of the Way. Their minds are one with the atmosphere of the fountain, the stone, the grass, and the tree, changing through the four seasons. This is the true manner in which those who are followers of the Way love mountains and rivers.[5]

In western Kyoto Muso developed the garden at Tenryû-ji at the foot of Arashiyama and the western hills. Like Saihô-ji, Tenryû-ji began as a paradise villa garden surrounding a pool, and was later augmented by the elements of the Zen dry garden style [1-4, 1-5].

About 1339, Musô also contributed to the remaking of nearby Saihô-ji. Faced with the existing garden and the limitations of its dense clay soil and nearly continuous shade cast by the adjacent mountain, Musô's strategy relied more on pruning and adjustment than on removal and new beginnings. The pond—its shape based on *kokoro*, the Japanese character for spirit or heart—remained the garden's central feature, its flow sufficiently restricted to create a slightly muddy composition that heightened the reflectivity of its surface [1-6]. Water was an central element of Japanese gardens tracing back to the Heian period (794–1184), a tenacious landscape feature that would reappear in full force in the stroll gardens of the later Edo period (1603–1867). The pond extended the perceived limits of the garden by doubling its image through literal reflection and shaping. Its irregular contour used inlets and islands and peninsulas to convolute the perimeter of the pond and further heighten the sense of distance and space. As one walks through the garden, its features are revealed one by one in an unpredictable sequence; never does one enounter the grand vista of the French tradition or the geometry of the Italian Renaissance garden. In their place individual features articulate the flowing carpet of moss. The detail is immediate; the whole elusive. The composite experience of Saihô-ji accumulates impressions and ultimately relies on memory and understanding; unlike the arrival of enlightenment, understanding the garden is not sudden.

[1-4]
TENRYÛ-JI. KYÔTO, JAPAN. CIRCA 1338. MUSÔ SOSEKI.
The pond garden with Arashiyama beyond.

[1-5]
TENRYÛ-JI.
The rockwork, in places, reflects the influence of painting from the Chinese Song dynasty.

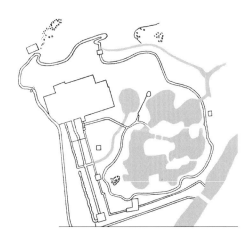

The idea of transition, so critical to Japanese garden design, also plays a central role at Kokedera—the moss temple—as Saihô-ji is familiarly known [1-7]. Moss dominates all parts of the landscape; it is the primary material for surfacing, blending, and transition. But these ideas of transition and passage reappear at larger and more apparent scales as well. The formalized entrance into the garden—later perfected at the garden of the Silver Pavilion in eastern Kyoto—uses a carefully orchestrated progression from the strictures of the exterior world to the more relaxed world of the interior garden [1-8, 1-9].[6] Entering through a roofed gate, the eye encounters a straight path of stone that turns abruptly to the right. Along the right side of the path a white clay wall with a dark tile cap directs our attention to a second turn in the distance. The banked earth on the left mutes the rigidity of the path, a dampening further effected by its moss covering—the first statement of the garden's principal material. As one moves inward, the straightness of the path remains while its definition is further softened by moss, earth, and trees that increase in prominence as the presence of the wall recedes.[7] Visitors today follow a different sequence, however, and enter the main garden through a gate set in the wall of the temple's central area. In older times, one might have entered the buildings of the temple first, viewing the garden from the verandah of the pavilion or moving into the garden directly. Unlike in the past, access is now direct.

The vegetal palette consists mostly of maple and pine and bamboo [1-10] and is kept purposefully minimal, its simplicity appealing to Musô's Zen beliefs of humility. Although visually spectacular in autumn hues, color itself plays a lesser role in Saihô-ji than, for example, in the imperial villas of Shûgaku-in and Sentô Gosho. A nearly continuous texture of maple and pine replace the splashes of color or the flashes of specimen trees; moss greens replace the raked sand or pruned shrub more common to the dry garden.

Throughout the gardens, moss subsumes every surface, and every surface has been tinted by the presence of moss. The north sides of trees glow fluorescent with lichen; the ground below them forms a literal carpet of moss. It is said that over forty varieties of moss coat the garden of Saihô-ji, most prominent among these a soft, dense, rice-green species that models hillsides and the banks of the pond into a myriad of tufts. As in many other Japanese gardens *sugigoke*, or cedar moss, is prominent. Its texture, if examined closely, appears to be a minute forest of cedar or cryptomeria trees, hence its name. At

[1-6]
SAIHÔ-JI.
Subsumed by moss, the shoreline reveals few stones, quite uncharacteristic of later gardens.

[1-7]
SAIHÔ-JI.
The restrained palette of maple, pine, and stone—but most of all, moss.

[1-8]
SAIHÔ-JI.
Plan. The pond, in the form of the Japanese character for *kokoro*—heart or spirit—is the garden's central feature.

a distance, its reddish tones tint slopes and beds as an airbrush spray might heighten the modeling of a curving surface. At the hottest times of year, should the rainfall be low, the mosses appear russet and moribund; during the wet months, however, from May through July, they absorb the welcome moisture and spring into luminescence. While the moss covers the ground and trees and renders them less distinct as discrete elements, it also increases their illusion of contour as sunlight rakes their surfaces. In many ways Saihô-ji is better sensed in diffuse light, however, when the landscape as a field emerges from the shadows; or facing the sun, when the delicate leaves of the maples are backlit and moss repels the glancing light.

The precise extent of Musô's work at Saihôji is not documented. Being of the Zen sect his work in the lower pond garden was probably one of simplification rather than elaboration, or elaboration through simpli-fication. The restricted planting palette may be his contribution, perhaps having distilled a more extravagant variety of plant materials when the garden conjured a sense of paradise in earlier years. Known to be to Musô's design are the stone formations of the upper garden higher on the hillside. A roofed wooden entry gate [1-11] marks the passage from the lower to upper—or perhaps better phrased, inner to outer —zones. A stair zigzags steeply up the slope, the rapid movement and change of direction contrasting noticeably with the easy gait of paths around the pond garden. A major stone grouping is set in the formation of a cascade, although no liquid ever flows abundantly over the rocks except with rain [1-12].

[1-9]
SAIHÔ-JI.
The formal entrance
path and tile-capped
boundary wall.

[1-10]
SAIHÔ-JI.
The northern bamboo
grove.

[1-11]
SAIHÔ-JI.
The entrance gate to
the upper garden and
the dry cascade.

A high mountain
   soars without
    a grain of dust
a waterfall
   plunges without
    a drop of water
Once or twice
   on an evening of moonlight
    in the wind
this man here
   has been happy
    playing the game that suited him.[8]

The composition is decidedly horizontal, in contrast to the upright thrust of the rock groups of Tenryû-ji that reflect the influence of Southern Song dynasty China (929–1127) [see 1-5]. This is a dry cascade, a configuration of rocks that suggests the presence of water without requiring the actual liquid itself—a design idea that would exert considerable influence on gardens thereafter.

The notion of suggestion is fundamental to Japanese art in general and Zen aesthetics in particular. If one understands the circumstances, silence can become more powerful than a shout. In its absence, the water is felt more strongly, for this rock composition requires the mind of the observer to complete the setting before him or her. The missing water in the dry cascade pairs with the still water in the pond below. This could be its source, not physically, but spiritually. In the pond all water is rendered as one; yûgen derives from this mental conflation of moss, water, vegetation, shade, and reflection.

[1-12]
SAIHÔ-JI.
The dry cascade of the upper garden.

[1-13]
SAIHÔ-JI.
The pond on a December afternoon.

> In this small hut
>   are worlds beyond number
> Living here alone
>   I have endless company
> Already I have
>   attained the essence
> How could I dare
>   to want for something higher.[9]

As a physical totality, the garden of Saihô-ji comprises two parts joined by a gate under a ubiquitous coverlet of moss. The sense of the over-all organization, however, is minor in comparison to the involvement with the immediate. The garden, with its lush green blanket, shares the effect of reading a novel by Marcel Proust. As in the French author's works, the design and formal structure of the garden as a whole are distant and indistinct. We remain, instead, aware of the moment and the immediate setting. The moss garden exerts a cen-tripetal force that diverts our attention from the garden as a whole and instead directs us to experience the fragment as a whole. The world beyond the garden wall seems very far away [1-13].

Originally published in the *Journal of Garden History*, Volume 9, Number 2, April–June 1989.

Notes

1 Makoto Ueda, *Literary and Art Theories in Japan*, Cleveland: The Press of Case Western Reserve University, 1967, p. 63. The entire fourth chapter, "Zeami: Imitation, Yûgen and Sublimity," like the rest of the book, is a concise and well-written explication of Japanese thought.

2 Ibid., p. 61.

3 Ibid., pp. 66–67.

4 The Jôdo sect was founded about 1175 by the priest Hônen, who lived from 1133–1212. It is said to be the second largest Buddhist sect in Japan today.

5 Musô Soseki, "On Gardens and the Way," in *Sun at Midnight: Musô Soseki, Poems and Sermons*, W.S. Merwin and Sôiku Shigematsu, translators, San Francisco: North Point Press, 1989, p. 163.

6 The influence of Jôdo Buddhism was long-lived and directly influenced the use of the image of silver and reference to the moon at Ginkaku-ji, circa 1480. The mother of the builder of the Silver Pavilion, Yoshimasa Ashikaga, is said to have been particularly fond of the garden of Saihô-ji.

7 Marc Treib and Ron Herman, *A Guide to the Gardens of Kyoto*, revised edition, Tokyo: Kodansha International, 2003, pp. 107–109.

8 Soseki, "Poem on Dry Mountain (A Zen Garden)," in *Sun at Midnight*, p. 32.

9 Soseki, "In This Small Hut," in ibid., p. 57.

[2]

Traces Upon the Land: The Formalistic Landscape

1979

Why does Stonehenge seem so special and powerful [2.1]? Why does this primitive, literal pile of stones, set out upon the Salisbury Plain continue to generate a powerful aura when we build structures hundreds of times its size, in thousands of places, all over the world? Still, Stonehenge remains a very special landscape, a unique place. One can marvel at the size and weight of its elements, the efforts of early humans to erect them, and even attempt to reconstruct or relive the ritual that instigated its birth. But all of these are *analytical inquiries* and have little to do with its actual presence, which is primarily an *emotional response.* For the moment, let us set aside inquiries into function or ritual or meaning. Stonehenge, I believe, does what it does, and is what it is, because it represents a delicate agreement or dialogue between the natural landform and the human hand. That dialogue possesses greater richness than a monologue by either made or natural forces alone, and of far greater interest and intrigue. That dialogue is the subject of this paper.[1]

[2-1]
STONEHENGE.
WILTSHIRE, ENGLAND.
THIRD CENTURY
B.C.E.+.

[2-2]
THE TATRA
MOUNTAINS.
NEAR ZAKOPANE,
POLAND.
The rectangular cut
in the forest.

The story opens some years ago, on a drive high in the Tatra Mountains [2-2].[2] The destination lay near the border between Poland and Czechoslovakia not too far from the mountain village of Zakopane. The Tatras are made of granite. They are steep and craggy, particularly as the road winds and climbs toward the summit and the small lake called Morskoe Oko, the "eye of the sea." Some slopes—depending on their incline, orientation, and hydrological factors—are covered by forests; evergreen forests that subsume the gray stone almost completely, save for some few outcroppings still to be seen now and again. It is a mountain landscape in a very fundamental sense. Rounding one curve, a small discernible patch of hillside came into view from across the valley. There, about halfway up the mountain, was a very precisely cut rectangle of clearing, cleanly distinguished from the fabric of the green pines. Below the rectangle, in a less clearly defined area at the foot of the slope, was a barn or some other field structure. The effect was immediate. Like an image from a slide projector hesitantly snapping into focus on a screen, the features, the colors, the actuality of the subject became apparent for the very first time.

That rectangle cut from nature, that strong statement of human geometry set in contrast to the apparent chaos of nature, somehow rendered the expanse of granite—and the vegetation that sheathed most of it—lucid. For the first time the mountain expanse actually became *real*. For the first time it had been translated into human terms. One could now relate to the landscape rather than just move through it, or merely look at it. The presence of that rectangle, though conceived simply as a way of clearing a field without aesthetic intention, gave meaning to the forest—at least to me. The rectangle was both a positive and a void simultaneously; an absence and a presence that defined the forest by what it was and at the same time what it was not.

For me, that enigmatic and paradoxical moment generated a very distinct presence. That experience, like others similar though different from it (Stonehenge, for example), has remained with me through almost ten years.[3] In the interval I have tried from time to time to understand the causes of such things; or more simply phrased: Why? This paper, however, presents more of a limited explanation than a definitive word on the matter. At this time I can propose no real answer; simply a proposition and some discussion of the considerations that bear on that premise.

The premise is a simple one. It is a premise or an observation rather than a principle, a truth, or a law, because it is first, personal, and second, unprovable.[4] The relationship of the natural to the constructed order raises some important questions that should be answered, and I do hope someone will answer them someday; but I am not that person nor is this the time for such an undertaking. My role here is in some ways like a scout, or perhaps a provocateur, offering some working hypotheses and perhaps instigating some thought on the part of the reader. The premise is simply that the co-presence of the so-called natural and the so-called man-made leads to a feeling or an aesthetic presence that is different from, and usually greater than, the presence of either of these others existing in isolation.[5] On the other hand, the by-product of the two ordering systems can be negative rather than positive; that is, the sum can be *less* than either one alone. It is a question of balance.

Thus the natural—as manifest in land forms—in simultaneous juxta-position with the constructed—represented, for example, by geometry —generate (at least for me) what might have been referred to by landscape romantics as the "sublime." Susan Langer, in *Feeling and Form*, referred to it as the "esthetic emotion."[6] I simply call it *presence* —the basis of aesthetic response. Unlike the romantics, I rarely find the feeling in nature alone, and even more rarely in architecture, except in the very best. But I find it quite often, though not often enough to be sure, in the combination of the natural and the geometric. Just why this occurs, I will tentatively examine in the course of this paper.

In addition to a discussion of the co-existence of the natural and the man-made orders, three other concerns present themselves. First, should there be a distinction between the man-made and the natural order? Is something man-made artificial or should we always consider mankind a part, though a rather special part, of the natural order of things? This in turn raises another question, the question of order. If humans are indeed a part of nature, a biological relationship few would question, how is human effort distinguished from the natural? The most outstanding differences seem to be length of duration and the scale of the ordering principles. Hence there will also be some discussion of the idea of order and how it bears on the subject at hand. The third and last concern addresses the nature of formalism, as found in the subtitle of the paper, "The Formalistic Landscape."

Order/Disorder

Many people prefer to pit the man-made and the natural in a state of opposition, as two distinct things, two distinct sets with no area of overlap. When regarding the ordering of the landscape, however, it might be more beneficial to think of them in another relationship: that the man-made and the natural occupy two points on the same scale. After all, man is an animal like the others; only certain specific char-acteristics distinguish him from other members of the kingdom.

Order, as used here, regards a systematic and perceivable way of establishing the relationship of one element to another, as in the order of things. Chaos, in contrast, is the absence of order. In an article entitled "Order and Complexity in Landscape Design," Rudolph Arnheim defined order as "the degree and kind of lawfulness governing the relations among the parts of an entity."[7] Most people could accept this. More unusual is his definition of disorder. He writes that "disorder is not the absence of all order but rather the clash of uncoordinated orders."[8] From this definition we can extrapolate that perhaps chaos is less the total absence of order than the manifestation of an order that cannot be visually perceived. An application of this idea (visually-unapparent order) is central to the points discussed here. For if we can say that everything is ordered, our problem would lie only in its perception. This definition is inclusive rather than exclusive and puts both order and chaos in the same field.

A sense of perceived order depends on both the scale of the inquiry and the experience of the observer [2-3]. Order is easily sensed at the micro-scale: we learn to find reason and logic in cellular structure viewed through a microscope. In turn we also sense order at the macro-scale: the patterns of cities and forests, for example, all clearly emerge as structured when seen from the air. Even Los Angeles, which represents another physical extreme of sorts, evinces a logic and order as a built environment in relationship to its physical setting, apparent when seen from a sufficiently elevated position.

Order is most elusive at the scale of the human being [2-4]. Here we encounter difficulty in detecting order and structure. When we walk

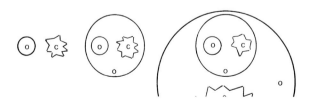

[2-3]
ORDER/CHAOS AND THE
SCALE OF INQUIRY.

through a forest, for example, we do not sense the system in the same way that we do when walking through an orchard. Were we to zoom back at a constant rate, however, at some height the order of the trees would begin to emerge. Not that geometrical spacing would appear; of course it would not. But there would be certain locations for that kind of tree, perhaps where sunlight fell; other species might cluster in closer proximity to a source of water. As we zoomed back farther still we would note that the vegetation of the western slope nearer the bay was far denser than that on the drier eastern hillside. As we continued our movement upward we would find that pockets of redwoods requiring increased moisture levels than other families thrive in bounded topographic depressions. Climbing still higher we would discover a similar logic in the growth of all coastal redwoods, and the ordering of trees and vegetation in the landscape would become clear. It would all seem to display sets of appropriate relationships.

We can discern from this example that we experience the greatest difficulty in perceiving order near the scale of the human body. But this should not imply that the constituent elements of the landscape are chaotic; it instead demonstrates that at least momentarily—or at certain scales or under certain conditions—we cannot distinguish the ordering system. The question then is not that the man-made and the natural are inherently distinct, but rather a question of the degree of readily perceptible order.

And humans seek order; it seems to be one of our basic activities.[9] We order a landscape to make it psychologically "comfortable." The conscious shaping of elements on the land, carved from a wilderness,

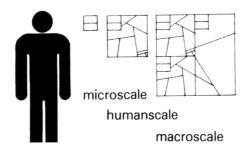

microscale

humanscale

macroscale

parallels the furnishing of an interior, to make an outdoor room comfortable on a very large scale. Ordering is an act of domestication. Ordering space is a way of reaching homeostasis in the environment —or to put it in Mircea Eliade's terms, it is the founding of "sacred" space. Only sacred space, in Eliade's view, is real and actual. Profane space, at least to the religious man, has no existence. In *The Sacred and the Profane* Eliade writes: "For religious man, this spatial homogeneity finds expression in the experience of an opposition between space that is sacred—the only *real* and *really* existing space—and all other space, the formless expanse surrounding it."[10] The rectangle cut into the forested Polish mountainside read as sacred space. The geometric order defining that void demonstrated an evident act of the human hand. Thus a landscape in a geometric order strikes us in a powerful way. It is potentially sacred in Eliade's terms, and it is good.

Wilderness, as the geographer Yi-Fu Tuan explained in *Topophilia*, is a rather recent concept in the West.[11] Historically, wilderness—as in wild—was a place to be feared; a place to be won; something to be controlled. For many centuries the city, as in *civis*—hence civilization —was *the* place to live. As some of its inhabitants soured on cities, they looked again to nature, no doubt to regain the divine Eden. They did not look to wild nature, however, but to a landscape whose order they could sense, or to a place that would comfort: to the garden, or the reshaped field or wood. As Tuan noted, the garden represented the comfortable interface between the city and the wilderness. Historically regarded with suspicion since the fall of the classical states (though a necessity in medieval times), an attitude exacerbated by the increasingly intolerable living conditions of the Industrial Revolution, the city was a place from which to escape. But the uncivilized and inhospitable, if at times glorious, alternative of the wilderness was hardly inviting. Only when secure in our knowledge of the natural forces and able to manage these forces, do we look again on nature as benign. As Gaston Bachelard quotes Henri Bosco in *The Poetics of Space*, "When the shelter is secure, the storm is good."[12] The garden, as the best of both worlds, is still hungrily sought in the move to suburbia.

Geometric landscapes

Certainly the geometric landscape represents a conscious and assertive human attitude toward the natural order. The seasons change, the years pass, plants and trees come and grow, but in the baroque pleasure garden, form continues relatively unscathed [2-5]. Yes, leaves fall, and their colors change, and the dormant parterres will come alive once again in the spring. But the order and the configuration remain ever the same. The topiary tree represents one of the supremely egocentric acts in the making of the human living environment [2-6]. Constant pruning and restricting of its branches thwart the natural forces by which a tree seeks its own contour. The form becomes architectonic, geometric, and ordered—but somewhat perversely executed in living material. One can almost feel the tension. Like an anxious dog pulling at the leash, the plant seems to want out. And like the lion tamer with his whip and chair, the gardener with his pruning shears calls the would-be offender back into line.

And yet the aesthetic presence of topiary can be astounding if we are able to suspend our moral sanctions. In a garden such as Versailles or Vaux le Vicomte the juxtaposition of the heroic scale of the architecture, the architectonic landscape with its topiary, the less-formal natural elements, and the more natural plantings, creates an environmental ensemble in which the lords may comfortably live untouched.

Essentially, Vaux le Vicomte constitutes a power play executed in plants and space. The vast axis that unifies the scheme leads from the moated chateau to the statue of Hercules in the distance and the heavens beyond [2-7]. Vaux was realized by the same design team employed by Louis XIV at Versailles some years later: André le Nôtre, Louis le Vau, and Charles le Brun. The scale of Vaux seems grand but within reason, within the grasp of the human mind. Its buildings and gardens hang together as a unit, unlike the gardens and park of Versailles where the sundry parts dissolve the sense of an entirety. The experience of Vaux is by no means boring, though the idea behind the garden is simple and strong. Details span the scales from the distant to the immediate, with events to enliven the paths between them. Trees may be cut architecturally in accord with the formal scheme most easily comprehended in plan, but the composition is neither two-dimensional nor static. Here Le Nôtre, with considerable assistance from nature, has really created a balanced masterpiece of grading, gardening, architecture, and artifice.

While some might say that the baroque garden too obviously represents the human control of nature (an actual dominance), we can safely say that over time nature, like the casino, always wins. Imagine a year or two without human care. Think of the changes that would occur to

this landscape and realize who really holds the trump card. Nature, it seems, has done little more than lease this space to the French, and the mortgage may be foreclosed at any time in the future, without prior notice. Payment must be made in care, and in the investigation and knowledge required to maintain the landscape.

And yet geometry is but one possible means of ordering. We might question why it has been used so extensively as a means of arranging landscapes, especially in the West. There are, no doubt, many good and philosophical reasons for the practice. Among these might be a series of ideological associations with geometric figures such as the square, particularly when related to the cardinal points; and the circle, which implies a central locus, a center of the world. There have been many eloquent expositors, including Thomas Jefferson on the square and John Brinckerhoff Jackson on the grid.[13] Eliade also discussed these at depth and they need not be restated here.

Geometry, when paired with symmetry, arguably represents *the* most human structural device. Bilateral symmetry appears frequently in nature, even in our own body form, but never to the extreme that it infuses man-made systems. In symmetry we deal with an abstraction, a pure distillation of geometry with no equivocation about intention. Though in contrast with most natural phenomena geometry is the most apparent, most easily perceived form of order. Geometry informs overtly straight lines and regular curves, shapes rarely if ever found in flowers and forests. There will be no problem of reading this place as one without the intervention of man, and no ambiguity as to its maker. Under the general heading of geometry could be included various subsections such as symmetry and repetition. Like musical theory, we often talk of rhythm in design: equally or purposely unequally spaced bays or columns, that unfold "rhythmically" as we move through them. Some landscapes are rhythmic and apparent, especially the French formal garden which pushes geometry, symmetry, and repetition to an extreme [2.8]. In some ways the highly formal geometric gardens are more like a straight antiphony, where many voices join in harmony about the same melodic line, like Gregorian chant. When we move to less apparently ordered gardens, such as the English landscape garden or the Japanese stroll garden, the rhythm and the interweaving of melodic lines, and the harmonies and the syncopations, become more complex. These gardens demand a greater involvement and discernment from the observer. Geometric orders are direct, or "fast" in the terminology of minimalist sculpture and painting [2-9].[14] Fast in

[2-5]
RAMBOUILLET. FRANCE.
CIRCA 1670.
CLAUDE DESGOTS.

[2-6]
VERSAILLES. FRANCE.
1661+.
ANDRÉ LE NÔTRE.
Topiary as an architectonic plant form, an intermediary between landscape and architecture.

[2.7]
VAUX LE VICOMTE.
FRANCE. 1656+.
ANDRÉ LE NÔTRE.
The principal axis.

that they can be read, at least as to their structure, almost immediately. Their complexity lies at another level. One grasps the overview easily; the richness evolves in the experience of the subsets. In naturalistic landscapes the elements unfold, often in a choreographed sequence, so that prior segments of experience must be held in mind. Meaning, and an understanding of the ordering system, come after the fact. Compare a vista of William Kent's work at Rousham Hall (1738) with the view from the terrace at Versailles, built some half a century earlier [2-10, 2-11].

The question of context

Any element, any bit or piece of anything, does not exist in isolation; it always relates to other bits and pieces of the same kind and those of other types. Meaning is not inherent in the single bit, but fixed by its relationship to others.[15] Thus an element acquires meaning or has its meaning defined only within a specific context. In language, a word is ultimately defined only when used and not while standing alone in a dictionary. In a landscape, it is the context provided by all the elements as a group and system that structures them; these interrelationships ultimately define the element.

While tribal societies normally utilized natural elements to construct their living compounds they often relied on geometry to arrange them. The real transformation occurred when natural elements took more permanent form on the land, however. Stonehenge was built, it is believed, as a sort of astronomical observatory, a place where humans sought an understanding of the universe and informed prediction. Approaching the rough megaliths on the broad Salisbury Plain, one senses something powerful and perhaps mystical. Although at first reading as an anachronism, some presence transforms this primitive pile into something timeless. Stone circles need not be as elaborate as Stonehenge, however. The circle of standing stones at Stenness in the Orkneys also demarcates a space, however loosely. The linear alignment of stones at Carnac in Brittany manifests the same intention to intervene and assert [2-12]. Here the now-informal stones, having been worn over the centuries, occupy a *pastoral* setting but positioned in a *geometric* order to conjure spaces, circular or linear. These spaces distinguish themselves in feeling and form from their surroundings. Not that they are completely separate from these surroundings; there is an empathy between them. Both groups are linked *and* independent: a "both/and" relationship. It is space with quotation marks around it, a dotted line; it is a virtual demarcation rather than concrete physicality.

The question of context and its relation to the perception of order is complex; examples of its misinterpretation are many. As a reaction to, or relief from, the pervasive rectilinear geometry in which most of us dwell, designers at times propose sculptural and curvilinear worlds. But the effect of the curve is easily diminished or lost when the entire work is executed along those lines; that is, employing only curvilinear shapes and forms. Juxtaposing the straight against the curved heightens and reinforces the presence of both, as each helps define the other [2-13].

The coexistence or juxtaposition of man-made and natural elements functions in a similar way. Their co-presence often produces a distinct experience more powerful than that produced by either ordering system standing alone. It would seem that this effect derives from the presence of one read against the presence of the other, in a form of dialogue. We focus on the man-made in reference to the natural, and vice versa. When we photograph something from close up, the depth of field of the image is limited: focus on a near element and the background is out of focus. Bring the background into sharp detail and we lose the foreground. Yet we are continuously aware of the presence of all elements, even those that are out of focus at any given time.

Simultaneous contrast pervaded the work of the Finnish architect Alvar Aalto [2-14]. In Aalto's architecture systems of order function more fully than as mere vocabularies of form. Aalto carefully played the rectilinear against the curved or the angular, each supported by its antagonist. When played against the straight line the curved wall seems more curved: it has a reference, a foil.[16] The straight line appears straighter than one of many in a completely rectilinear situation [2-15]. Thus, essences can be defined by what they are not, in addition to what they are, and a two-part definition or reference system seems to heighten the experience of a place. As in a mirror, each system finds its reverse image in the other.

Formalism in intention and perception

It seems appropriate here to interject the third subject that bears on the subject under discussion: that of formalism. Before proceeding any farther it might be helpful, or indeed critical, to derive a working

[2-14]
HELSINKI TECHNICAL UNIVERSITY. OTANIEMI, FINLAND. 1964. ALVAR AALTO. The raked and curving forms of the lecture hall/amphitheatre/ reception hall articulate the rectilinear geometry of the classroom and laboratory wings.

[2-15]
FINLANDIA HALL. HELSINKI, FINLAND. 1971. ALVAR AALTO. Contrasting formal idioms in section as in plan.

[2-16]
VILLANDRY. FRANCE. SIXTEENTH CENTURY. REMADE TWENTIETH CENTURY.

[2-17]
IDEA/FORM IN THE NATURALISTIC (UNKEMPT NATURE), GEOMETRIC (AS IN THE FRENCH BAROQUE), AND PICTURESQUE (IMAGE OF NATURE) GARDENS.

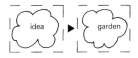

Geometric

Picturesque

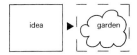

definition of formalism. The dictionary defines it thus, "strict attention to forms and customs, as in art or religion."[17] But the dictionary usage, though informative, does not neatly coincide with an appropriate definition of formalism as applied to the design of environments. Used within the context of this paper, then, formalism should be taken in this way: a conscious regulation of human activity that usually (though not always) results in a distinctly perceivable order or form. Admittedly, this is a limited definition and perhaps only valid within the present context. Sadly, the scale with which we measure formalism might appear circular rather than linear, depending on our use of the term.

Though a working definition of formalism is necessary to continue any analysis of the landscape, it is more important to distinguish between formalism of intention or instigation, and formalism in the final shape of a landscape; or to put it more succinctly, whether the formalism is found in the concept or in the conception. Thus, when looking at Villandry, we *see* a garden that is highly formalistic [2-16]. The concept is a simple one: the management of nature to such a degree that it will grow in the intended form, that is, in apparent constructed order called geometric. But what of the naturalistic landscapes in England or its relative in Japan, the stroll garden? Here the landscape appears highly naturalistic. It looks like nature; there are no tell-tale signs of human intrusion to tip its hand, as geometry does in the baroque garden. But is it really any less formalistic in *concept*?

I would suggest that these naturalistic gardens are not only equally formalistic, but actually more formalistic, at least in ideas [2-17]. Here we not only demonstrate that we are able to channel nature into a new order, but that understanding nature so well we can coerce it or even recreate it in its own image. Thus the relation of the concept and the conception in formalistic terms is almost paradoxical: what seems most formal when seen is less formal in conception, and vice versa. This is not to make value judgments, though today's Californians always seem to do so: if a space is geometric, if it has hard surfaces and is not green, it is "bad" because it is not "natural." Wilderness on the other hand is "good." And yet as Martin Krieger, like Tuan and others, has pointed out wilderness is an urban creation.[18] It is only the surety of control over nature, and the power to absolutely destroy it, that provides our comfort in the world—not in a garden, but on a global scale—and allows us the conceit that wilderness is good. We have reached the point where wilderness is our garden, tame and circumscribed, an endangered species that we keep under control like a pet. Even

Niagara Falls was stopped, reinforced, and rebuilt to maintain its natural "wild" shape for all to see.[19]

How formal or geometric can we get? From France we move to Japan. Let us examine two gardens in Kyoto, first the robust rock and gravel garden of Ryôan-ji, and thereafter, the upper stroll garden at Shûgaku-in. Ryôan-ji dates from the end of the fifteenth century [2-18]; Shûgaku-in almost two hundred years later [2-19]. Ryôan-ji is a garden of limited area focused on itself as a place for contemplation. It is a psychologically boundless space physically enclosed by a circumferential wall. It seems austere and abstract, and to the Westerner it is hardly a garden at all. Shûgaku-in, built as an imperial villa, is a garden of a far different stripe. Compared to Vaux, Shûgaku-in is also confined, though not to the same extent as Ryôan-ji, nor in the same way. At least in Shûgaku-in we encounter masses of plantings and trees, and to the eye the impression is more "natural." But is it? Using the system of formalistic classification outlined above, Shûgaku-in, as a representative of the seventeenth-century Japanese stroll garden, could be taken as more formalistic than the abstracted rock garden; if so, it represents a very presumptive and human act—at least in principle if not in scope [2-20, 2-21].

Look at it this way. As a *kare-sansui,* or "dry garden," Ryôan-ji reduces its palette to stones or sand and a bit of moss. The fifteen stones carefully arranged in five groupings are never all seen within the same view. The gravel is raked and the aesthetic is *overtly* abstract and symbolic. The human intervention is obvious; it is almost architectural. But Shûgaku-in is equally a human artifact. The pond is artificial, the result of damming the side of the mountain to capture seepage and rainwater; an impressive feat for the time. All the trees are planted, everything has been composed and arranged.

Vaux says: "Here is nature in a human order. Obviously this garden has been built and maintained."

Ryôan-ji says: "Again the human has created here. We have selected elements from nature to serve symbolically as representative of *all* nature. And we maintain this constructed landscape."

Shûgaku-in says: "Look, Nature!" or "We understand nature so well it need not look *as if* it's been made at all."

The human mind now understands not only horticulture and natural forms, but also the ordering principles behind nature, and can now

Intention/perception in the French and Japanese garden of the seventeenth century

	Intention	Perception
Vaux	formal	formal
Shugaku-in	formal	natural

A comparison of the order and elements in three gardens

	Elements	Order
Vaux-	n	m
Ryoan-ji	'n'	m
Shugaku-in	n	'n'

Relations of elements and order in the landscape

	Nature	Alignments	Geometric garden	Garden ruins	Architecture	Picturesque
Elements	informal	'informal'	informal	formal	formal	informal
Order	informal	formal	formal	'informal'	formal	'informal'

recreate nature in its own form, selectively, and even *improve* upon it. In effect, we construct a meta-nature that possesses distilled essences, like a perfume that extracts, concentrates, and blends natural fragrances to a hyper-natural degree [2-22].

Aesthetically-unintended dialogue

An aesthetic presence in the landscape may result without an aesthetic intention. Settings around the world illustrate the juxtaposition of human and natural orders from which presence may result although never intended by their makers. The straight lines of jetties contrast with the curve of a shore or with the seemingly random quality of the materials of which they are composed [2-23]. Or hedgerows—walls of living vegetation that articulate and subdivide the land—produce a strong aesthetic effect although their purpose is strictly functional. Hedgerows reduce the velocity of the wind across otherwise treeless plains in places such as western Japan or Jutland in Denmark [2-24]. The straight lines in plan, when fitted to the gentle undulations of the landscape at its roots, deform and assume a softly modeled form. Like the gentle rhythm of the rise and fall of the telephone lines strung in graceful catenary curves between the upright poles across the American landscape, the lines of trees define the terrain. Forms of hay, whether stacks or fences, also exemplify the practice of putting natural material into an order derived from process [2-25].

Whenever and wherever, human existence leaves marks upon the land. These marks can be divided into two basic types: trace and intent.

Trace is the record of man's efforts and actions without a conscious awareness of their imprint. This category includes such phenomena as dirt, grease spots, signs of use and wear, or the trampled grass of a footpath.

Intent is the mark of man's conscious attempt to intervene in the environment in some way. This intent need not be aesthetic; it may be purely functional, as illustrated by the haystacks and fences mentioned above. A functional intent may produce aesthetic byproducts, however. This notion—that aesthetic consequences may even derive from actions that are seemingly non-aesthetic—is central to landscape typologies like the hedgerow, and particularly critical to the under-standing of land art.

2-18]
RYÔAN-JI. KYOTO, JAPAN.
FIFTEENTH CENTURY.
The confined garden as an infinite landscape.

[2-19]
SHÛGAKU-IN VILLA,
KYOTO, JAPAN.
SEVENTEENTH CENTURY.
From the upper garden, a vast panorama of borrowed scenery in the natural order.

[2-20]
INTENTION/PERCEPTION
IN SEVENTEENTH-
CENTURY FRENCH AND
JAPANESE GARDENS.

[2-21]
ELEMENTS AND ORDER
IN THREE GARDENS.

[2.22]
RELATIONS OF ELEMENTS
AND ORDER IN THE
LANDSCAPE.

[2-23]
JETTIES.
BRITTANY, FRANCE.
The presence of the
unintended aesthetic
gesture.

[2-24]
HEDGEROWS.
JUTLAND, DENMARK.
Straight lines deformed
by the contour of the
land.

[2-25]
HAY-DRYING FENCES.
EIDSBORG, NORWAY.
Linear compositions
marking the hillside.

Traces upon the land

The co-existence of the natural and the constructed orders suggests considering constructions in the landscape as marks of human intent. The work of two sculptors seem of particular relevance here: Michael Heizer, one of the first artists to deal with the land on a large scale; and Robert Smithson, whose works include the *Spiral Jetty* in / on the Great Salt Lake; an artist whose artistic stance accepted that all systems of order, including nature itself, are affected by entropy; all systems decay and move toward stasis.

There are essentially two ways in which sculpture may relate to the land, as John Beardsley noted in his catalogue for the exhibition "Probing the Earth: Contemporary Land Projects," shown 1977–1978 at the Hirshorn Museum in Washington.[20] The first, the traditional setting for sculpture, provides a context or a site for the work; the landscape is the space that a piece occupies, at least in part. Of course, a sculpture always co-exists with that space; in a single instant it both inflects the properties of the space and is itself inflected by it. The obelisk in St. Peter's Square in Rome serves as a marker for the converging foci of the ovular plan. In turn, its significance as a cultural symbol as well as a spatial marker derives from the particular context of the piazza, which lies in Rome and *not* in Egypt [2-26]. At the physical or perceptual level it is a point within a defined urban void. As a cultural or literary marker, it recalls another point in space at another place on the globe, quite distant from the shores of the Holy See.

But sculpture may address the land in a second way, and that is to engage it more actively. This engagement is critical to understanding the large-scale art projects referred to as "earthworks." Here we do not regard the piece as a distinct unit set in a landscape. It is site-specific. Its relation to the land is integral. It uses the very same stuff of the land itself, and it is often only the ordering system that distinguishes the art contrived from the untouched land in and around it. Once again our basic acknowledgment of the piece and the source of its effect upon us derives from the contrast in ordering systems, or more precisely, a contrast in the scale of the ordering systems.

Works by these two artists appear to me as being extremely powerful. They must be powerful in actuality though I have seen none of them.[21] They must be powerful in actuality because they are powerful even in photographs, but of course that is no guarantee.[22] As noted earlier, however, systems of order may be identifiable at high altitude that are not sensed by a human being walking on the ground. The works share strong generating ideas that are in themselves "high altitude" concepts; but these are complemented by strong "ground" ideas, comprehended as we move through them.

Perhaps there are always two landscapes: one that we physically perceive and one that we mentally construct. We might say that successful earthworks generate a presence at both levels—as we read it from above or think through it as an idea; or as we actually encounter it, move through it, or even predict the experience. On the other hand, on site we do not simply experience fragments of a totality whose synthesis and order evade us at ground level. In the tradition of the English landscape garden, we sense the totality as well as the parts. A dialectic emerges: the play between the micro- and the macro-orders, and in turn, the play between each of these orders in relation to the physical topography itself. Thus our perceptions instigate twin and simultaneous communication cycles, while testing the relationship of the actual and the construct. These are rapid oscillations, perhaps at a speed approaching that of alternating electric current, and the oscillations are sensed as a single continuous perception, as a unity.

Some specific examples may help clarify the point. The Nazca lines were made some hundreds of years ago by a people no one is sure of, for a purpose that remains somewhat unclear.[23] The lines stretch for miles. They have no apparent terminal points. From the ground they are hardly distinguishable from the land into which they have been incribed. Robert Morris, a sculptor who journeyed to Peru, described them in a telegraphic manner in *Artforum*: "The horizontal becomes visible through extension. ...The further down the line one looks the

greater its definition. Yet the greater the distance, the less definition of detail…The Gestalt becomes stronger as the detail becomes weaker."[24]

The edges of the lines are weakly defined and thus the lines are hard to read from the ground. From the air they become more evident. At ground level we might easily dismiss them as a mirage, like seeing faces in a cloud formation. The lines seem to be there, but are they? As we walk, the slow speed of our travel retards an integrated reception of the information. From the air there is little to dispute that these lines are in fact fabricated; their form, if not the intent, is clear.

Michael Heizer's *Double Negative* (1970), executed near Overton, Nevada, may be the most potent of the earthworks [2-27].[25] It was also one of the earliest efforts to move earth on a large scale as an artistic activity. Heizer made two sloping cuts into the edge of a mesa, one on either side of a small arroyo. One descends the slope into the canyon created by the cut, a topographic maneuver that allowed the bulldozer to do its work. As we descend, the sides rise. As the sides rise, our view becomes constrained between earthen blinders some twenty feet high. We can examine the rich color of the ground, the strata that tell, in Heizer's terms, "the past and future of the mesa."[26] But we are focused ahead, across the rough edge of natural erosion to the corresponding void on the opposite edge of the gully. The material is consistently the same throughout the work and the landscape itself. We move from a flat plane to a channel defined by that very same material; what was originally only the ground plane of the mesa is now the wall material as well. Our focus falls not on the material, the solidity, but instead upon the invisible, hypothetically defined spatial channel that links us to a similar though corresponding condition on the opposite side. Again we are

forced, or at least coaxed, into constructing a mental landscape. Perhaps it is always this play between the mind and the physical world that creates presence.

Smithson's *Spiral Jetty* (1970) operates in a different manner. In many ways the jetty's form relates to that of a labyrinth, although there is obviously only one way we can trespass upon the path. The work possesses many dimensions nonetheless. One is the notion of closure and opening, since a spiral can effect either, depending on the direction of movement. There is also a second spiral, of water, in addition to the one made of black basalt, limestone, and earth. Smithson's project consciously constructs a discernible order in the landscape: "A bleached and fractured world surrounds the artist. To organize this mess of corrosion into patterns, grids and subdivision is an aesthetic process that has scarcely been touched."[27]

We move toward the center and our pace quickens; but nothing fills in the center, or at least nothing that seems special. The relation to the labyrinth lies in the concept behind the form; our experience and our concept differ completely. In form, the order is clear. In actuality we are aware of the linearity of the path, what lies ahead, and what we have passed. We look to either side as a confirmation, a fix in space. But this linearity contrasts strongly with the field or circularity of the idea of the spiral. The path makes us look down and then look up. At certain points our view falls on the lake, and at other times back on the land. The experience parallels in some ways the zigzag bridges of the Japanese garden that spurn the continuous axis with its one-shot experience; on an axis, movement serves only to confirm. Instead, by directing our view first to the left, now to the right, all views remain distinct and fresh: not a single vista, but a multitude. And by controlling the placement and the varied finishes of the stepping stones, the visitor is forced to look down for grounding and then upward to a new vista. Similar things happened on the jetty. Aspects of similar ideas also informed others of Smithson's works, in particular the *Amarillo Ramp* (1973), which enfolded a progression in height as well as space, and the poetic *Broken Circle* (1971) in which he used, as did Heizer, the opposition of positive and the negative to define through contrast.

Some find it ironic that the jetty was submerged much of the year and barely visible beneath the surface of the Great Salt Lake. Smithson probably would have accepted it, however, since he willingly acknowledged, and even embraced, the idea that entropy was a part of all systems, including governing artistic practice. The jetty, as Smithson said, is "going nowhere, coming from nowhere."[28] Beardsley provides an excellent summary of Smithson's art worth quoting at length:

> Smithson had perhaps the most involved attitude toward nature, and one which best reconciled the human and natural elements that often make up our sense of a landscape. He combined a remarkable sensitivity to natural materials and unusual landscapes with the use of open, active forms such as the spiral, which imply impermanence and irreversibility. He was able to suggest through this combination the duality that he perceived in nature: the natural forces that formed materials would lead to their eventual erosion and disappearance. Smithson accepted signs of inevitable entropic change. He incorporated this awareness of entropy into his art by a choice of barren or disrupted sites, and the use of forms which implied change or increasing disorder. He rooted his art in the paradox that nature at once both creates and destroys.[29]

Order and entropy represent two poles on a scale. In the end, or so say the physicists, entropy will always win. Ultimately, no matter how sophisticated the system of order, it will deteriorate in time. For at best, nature gives the gardener a reprieve and the garden a stay of execution. The lion tamer, the gardener with his pruning shears, can only gain some time through constant maintenance. He must be vigilant and constantly aware, and on his guard. The lion—in this case the natural order—is always ready to jump, always ready to reclaim the land from the intruding hand. A forest's second growth is the scar on the tamer's hands, the climax forest that may follow the ultimate domination of things natural over those human. As in baseball, when it is the top of the ninth inning, the land is ordered, the human presence apparent. It seems all the forces are on the human side, with cities of concrete and steel and asphalt creeping everywhere. The trees stand neatly in rows, possibly even clipped in topiary form. The game may seem to be over. But we must keep in mind that like the home team in baseball, nature always bats last.

Originally published in *Architectural Association Quarterly*, Volume 11, Number 4, 1979.

Notes

1 *This essay is the earliest of the writings included in this collection and the one about which I have the most doubts. Today I would have to further qualify this assertion, since I have found it rather impossible to separate completely the syntactic and semantic aspects of perception. And as a whole I would also have to admit that some of these assertions are weakly argued and perhaps a bit naive. I remain interested in the observations, however.*

2 *Actually, the drive took place in 1969, today almost 35 years ago.*

3 *As noted above, far more than ten years have passed.*

4 *Perception and interpretation are so affected—some might say channeled—by culture and education, in addition to personal experience, that it would be almost impossible to ascertain where one ended and the next began.*

5 *This paper was written long before gender-free phrasing entered the scene and the term man-made was used throughout the paper. I have modified these transgressions where the same sentence structure could be substantially maintained. Others I have left as originally published. Today, of course, I would do it differently and probably use "constructed" wherever possible.*

6 Susan K. Langer, *Feeling and Form*, New York: Charles Scribner's, 1953.

7 Rudolph Arnheim, "Order and Complexity in Landscape Design," in Paul G. Kuntz, ed., *The Concept of Order*, Seattle, WA: University of Washington Press, 1968, p. 153.

8 Ibid.

9 *I still believe that we seek order in some form, although trips to some friends' apartments or to architectural studios at universities might undermine my assertion.*

10 Mircea Eliade, *The Sacred and the Profane*, New York: Harcourt, Brace and World, 1959, p. 20.

11 Yi-Fu Tuan, *Topophilia*, Englewood Cliffs, NJ: Prentice-Hall, 1974.

12 Gaston Bachelard, *The Poetics of Space*, New York: Beacon Books, 1969, p. 39.

13 See John Brinckerhoff Jackson, *Landscapes*, Amherst, Mass.: University of Massachusetts Press, 1970.

14 "Fast" in this context implies a direct reading of a work of art; simply; quickly. See Gregory Battcock, *Minimal Art: A Critical Anthology*, New York: E. P. Dutton, 1968.

15 *A later essay, "Must Landscapes Mean?"—reprinted in this volume—developed this idea at greater depth.*

16 This characteristic of Alvar Aalto's work is readily seen in such projects as the 1966 library at Rovaniemi or his Kulttuuritalo in Helsinki, 1956.

17 *Webster's New World Dictionary*, New York: Collins World, 1975.

18 Martin Krieger, "What's Wrong with Plastic Trees?," Working Paper, Institute for Urban and Regional Design, University of California, Berkeley, 1972.

19 *In the 1960s concern over the rapid erosion of the American Niagara Falls instigated the forming of a committee to propose policy for their preservation. Garrett Eckbo served on the panel. Thereafter, the flow of water was modulated by means of upriver dams; to facilitate stabilization work on the rock structure, at one point the Niagara River was constrained to one small area and restricted to barely a trickle—the falls were essentially dry.*

20 John Beardsley, *Probing the Earth*, Washington, DC: Hirshorn Museum, 1977, pp. 9–11.

21 *Over the years since this essay was published I have visited a number of these earthworks—but not the* Spiral Jetty, *unfortunately. Until recently it was inundated by the risen waters of the lake. The powerful conceptual gesture behind* Double Negative, *however, was still apparent even in its somewhat eroded state in 1999.*

22 *Alas, experience has demonstrated that a number of landscapes I have visited are far more impressive in photographs than in reality. The frame of the camera, the ability to wait for the perfect environmental conditions, and the photographer's own skill and creativity can greatly enhance the characteristics of the actual design. This is one thought I would dismiss or greatly qualify today.*

23 *Considerable work has been done on the Nazca lines since the mid-1970s when this paper was written, and the answers to many of these questions have been ascertained.*

24 Robert Morris, "Aligned with Nazca," *Artforum*, October 1975, p. 31.

25 *This statement, too, would need to be qualified based on subsequent experience.*

26 Michael Heizer, unidentified source.

27 Robert Smithson, "A Sedimentation of the Mind," *Artforum*, December 1967, p. 45.

28 Robert Smithson, in Beardsley, *Probing the Earth*, p. 86; actually a quote from Samuel Beckett.

29 Beardsley, *Probing the Earth*, p. 25.

3

Inflected Landscapes

1984

Even those architects most committed to the modernist manner have admired the quality of anonymous folk architecture. Though their evaluations might have derived primarily from building shape or the appropriateness of materials and construction technology, the relation of folk buildings and villages to their sites has rarely failed to elicit appreciation. Yet now, late in the twentieth century, the isolated structure standing in contrast to its landscape backdrop, with only the comfort of some transitional planting, represents the most common approach to siting the detached building. Were we to reconsider the range of possible building/topography relationships, however, we might realize that the object, dissimilar to the landscape, occupies but one of three categories. The other two, often more harmonious, are the *merger* of building to site and/or the *inflected* landscape.

In spite of extended discussion of the "order of nature," that order, at least in visual terms, is difficult to apprehend. The ordinary person, for example, finds little apparent pattern in the depth of the forest. The trees seem irregularly disposed; the mix of plant species belies no grand plan. From a vantage point sufficiently distant from the forest floor, however—high above in an aerial view or one afforded by a neighboring hill—an indisputable sense of order in the landscape becomes apparent. Trees are seen to grow primarily in the narrow valleys or ravines where the excess runoff of the rains or underground seepage provides more reliable sources of water. Certain species thrive in these conditions, perhaps growing on just one side of the hill where the sun, wind, and moisture are most favorable to their existence. Smaller plants occupy analogously appropriate locations on the land, deferring to the predominance of the trees and directed by other environmental factors. An order to the natural environment does, indeed, exist; but it is an order that may be understood better in mental rather than perceptual and biological rather than visual, terms. Only with sufficient distance, or, conversely with the extreme close-up of the microscope, is scrutiny rewarded with comprehension.

Industrial society tends to favor strong statements on the face of the land whether by intention or benign lack of consideration. Because the sense of order is most difficult to perceive near the human scale, those making places have devised effective strategies for making order manifest. One approach dismisses non-contributing elements, fashioning gardens or sacred spaces by *omission*. The dry gardens of Japan, produced under the influence of Zen Buddhism within a matrix provided by the Shinto religion, evoke an aesthetic presence by actions

[3-1]
DAISEN-IN, KYOTO,
JAPAN. CIRCA 1509+.
The landscape abstracted
and compressed; infinity
within the space of several
feet.

[3-2
VAUX-LE-VICOMTE,
FRANCE. CIRCA 1660.
ANDRÉ LE NÔTRE,
The garden of
restatement: all
elements refer to
and augment the
presence of the axis.

such as clearing a plane within a forest, modeling topography, or arranging rocks as the subject within a contrived void. By reducing the garden elements to rock, gravel, and a few bits of moss, a garden in Kyoto such as Daisen-in eliminates extraneous distraction and focuses attention internally [3-1].

Restatement exemplifies a second category: the repetition of related forms and/or their alignments within an explicit ordering system, designed so that each fragment restates and contributes to the power of the whole. The line and the geometric figure are vehicles commonly employed to render this repetition transparent. The French formal garden, perfected in the seventeenth century by the landscape architect André le Nôtre, is the principal representative of the type. The axis provides the overriding structure to which each element refers, analogous to the bass line of three-part baroque musical composition. At the estate of Vaux le Vicomte, for example, le Nôtre extended the primary axis from the oval salon of the chateau to the skies beyond the horizon [3-2]. The geometrically rendered plant materials, clipped in a severe topiary nearest the chateau, reinforced the architectonic concept of linear order. The treatment of the landscape became less mannered and more naturalistic as one moved away from the central line, eventually dissipating within the plantations of the surrounding woods.

In both instances the notion of garden implicitly constrains the relationship of the building to the landscape. At Daisen-in the garden exists only to be viewed: a trapped field of gravel and lithic composition, encircled by a plastered earthen wall. The space remains untrammeled,

witnessed only from the verandah that serves as a platform for meditation. At Vaux le Vicomte the garden and chateau are also firmly integrated, disposed along a single axis that weds natural to constructed elements. In both examples, however, the building stands freely on its site, perceived as a discrete object.

In the past architects, builders, and landscape gardeners have also attempted to create picturesque or naturalistic landscapes and even naturalistic buildings. Gardens intended to replicate nature are more easily realized, and are found in both the Eastern and the Western traditions. The merging of structure with landscape, on the other hand, has rarely been achieved to an ideal degree. Works such as Frank Lloyd Wright's Kaufmann residence, Fallingwater, or the Parc Güell by Antonio Gaudí, interest us precisely because of their inability to achieve complete mimesis with the natural terrain—as if the very imperfection of the merger produces the basis of its success in terms of perceived order.

It is doubtful that architectural means can truly replicate a natural order. The act of construction as the creation of places for habitation, work, worship, or pleasure, almost by definition distinguishes these places from the preexisting order. Fallingwater, in spite of the prevalent myth, hardly coincides with its surrounding landscape [3-3]. The cream-colored horizontal planes of the composition easily distinguish the house from the tone and form of the surrounding woods. As a balanced assembly, the mass of the central fireplace counters the sweep of the horizontal terrace-planes, locking the composition in equilibrium.

The vertical masses, however, are constructed of field stone laid in distinctly horizontal courses, as if to diminish the disparity between intention and realization. In Wright's work, the horizontal line of the stone or brick courses often counters the inherent verticality of the wall planes. The architectural composition addresses architectonic necessity; the selection of building materials, on the other hand, supports the joining of house and site.

A second example, the Parc Güell in Barcelona, was conceived as the matrix and first phase of a housing estate that was never completed. Built to a design by Antonio Gaudí and constructed by 1914, the meandering features of the park develop from the topography and native vegetation but would never be mistaken for untouched terrain. Even the most rustic of the park's constructions—the stone retaining-wall with its crude caryatids—must be regarded as an inflection of the land rather than a recreation of its prior state [3-4]. Indeed, the aesthetic power of the park—aside from its formal and chromatic beauty—lies in this *departure* from its prior state. Inflected landscapes such as this occupy that middle zone between the natural and the made, instigating a state of soft tension that evokes a perceptual ambiguity between the identities of both the natural and the constructed. Contributing to this ambiguity, the exact nature of their interrelation-ship seems to shift and change under differing environmental or temporal conditions.

A gradient of possible construction/natural site relationships could situate *merger* or coincidence (at least as an aspiration) at one end, and *distinction* at the opposite. Certain heroic modernist structures such as Le Corbusier's Villa Savoye (1929) might serve as extreme examples of the latter category, although even in this striking instance the distinction of building from site is not absolute [3-5, 3-6]. The roof garden, the siting of the house, and even the rigid block geometry of the villa reflect on the landscape through contrast and contradiction. Definition derives from opposition: what the structure is not, as well as what it is. Merger, on the other hand, is most clearly represented by the English landscape garden with the contrived naturalistic forms of its park blended into the wooded surrounds. Constructed caves or follies provide more architectural examples of this manner, as do the more contemporary energy-concerned structures with their "bermed" or buried constructions. Somewhere between the two extremes, though closer to merger, lies the landscape of inflection: places that retain in part the natural order or indigenous materials

[3-4]
PARC GÜELL,
BARCELONA, SPAIN.
1914. ANTONIO GAUDÍ.
The natural materials of the retaining wall and the caryatid soften the regularity of the order but hardly conceal it.

[3-5]
VILLA SAVOYE.
POISSY SUR SEINE,
FRANCE, 1930. LE
CORBUSIER.
The machine in the garden.

[3.6]
VILLA SAVOYE.
The window opening in the screen wall of the roof terraces frames the view and engages the landscape—at least visually.

while articulating an arrangement distinct from the prior form of the landscape—a distinction sufficient to generate a sense of a new entity.

[…]

Released from the burdens of function, earthworks are intended to be approached as art, and the ritual for their perception is preordained. Functional places for human habitation are another matter, although as strictly formal compositions both share common properties. The architecture of the Anasazi, built mostly from the ninth to twelfth centuries in the Four Corners area of the American Southwest, closely parallels the mood of the earthwork sculptures. For the cliff dwellings of Mesa Verde, Colorado, the caves created by seepage and erosion provided the superstructure, the frame into which each piece of the settlement was fit, each part adjusted to the whole [3-7]. Here, too, dwelling and mesa share a common material; but in the act of piling, chinking, and plastering the builders fashioned regularly shaped dwellings within the irregular confines of the mesa's cavities. Lacking caves suitable for habitation the Anasazi in Chaco Canyon, New Mexico, relied on a formally conceived plan when constructing Pueblo Bonito [3-8]. The arc of the layout developed over three major periods; and only in its mature phase did it complete the D-shaped composition of multistory apartments and round subterranean *kivas* that loosely face south. The construction materials remained those of the mesa.

The issue of appropriateness of building to topography, of merger or contradiction, informs this discussion. Yet the very distinction of

[3-7]
CLIFF PALACE,
MESA VERDE,
COLORADO.
ELEVENTH TO
THIRTEENTH
CENTURIES.
ANASAZI CULTURE.
The limits of the cave
structure the ordering
of the individual cells.

morphological orders, the *differences* rather than the similarities, I would suggest, are the sources of their vigor. Imagine in the place of these structures, adjacent to the true escarpments, a naturalistic construction of a cliff executed in sprayed concrete. If the imitation were so perfect that it might be mistaken for its prototype, the replica would still lack any sense of the historical record that infuses the natural strata. Would this imitation evoke the same sense of suitability or even awe produced by these native constructions—even in ruins?

But these dwellings are part of history. Technological and ecological necessity constrained their construction, restricting the builders to utilize the very materials of the site. What, then, might be parallel case studies from the modern era, in which the complexity of the twentieth century vastly complicates the problems at hand? There are two places in Scandinavia—of necessity in at least *semi*-rural settings—that provide instances from which we might learn. In the first, the extension to the South Stockholm Cemetery, the buildings and landscape stand in balanced articulation and in so doing define both the architecture of nature and the nature of architecture. In certain places the buildings configure a literal edge to the forest—both distinct from the vegetation yet curiously akin to the trees. At the Villa Mairea, the second example, the relationship changes from a literal to an analogical or even metaphorical one. The structure of the house rationalizes and reinterprets the structure of the trees in the forest. A romantic appropriateness results, with the interior and its structural system maintaining a continual dialogue with the surrounding pines.

[3-8]
PUEBLO BONITO,
CHACO CANYON,
NEW MEXICO.
ELEVENTH TO
THIRTEENTH
CENTURIES.
ANASAZI CULTURE.
The loosely south-facing
D form of the grouping
integrates the living cells
and round subterranean
kivas.

Erik Gunnar Asplund occupies a pivotal position in the history of Swedish architecture. As it had fallen to Alvar Aalto and Erik Bryggman in Finland, the task of "converting" the Swedish population to an acceptance of modern architecture fell to Asplund, Sven Markelius, Gregor Paulsson, and several other architects and theorists writing and building around 1930. Today Asplund is remembered primarily for three major projects that crossed from modernism to classicism, and ultimately to a unique blend of the two. The first design, the 1929 Stockholm City Library, fused prismatic and classical forms with the elegance of the modern idiom. Asplund also served as chief architect for the 1930 Stockholm Exposition, the first large-scale flowering of modernism in the North, where the new and the contemporaneous ruled unchallenged. And third, Asplund's work—with collaborator Sigurd Lewerentz—at the Woodland Cemetery exemplified his ability to conceive and execute a narrative landscape of trees and stone.

As early as 1915 Asplund had joined with Lewerentz to design the cemetery and chapels for Skogskyrkogården, usually translated as the Woodland Cemetery. The buildings that he and Lewerentz built there, and the entire setting, embody a superb use of contrasting orders. The preexisting landscape of the site was covered with some stretches of pine forests, and a portion had been cleared and used as a gravel pit. The first phase of the post-competition development (1918–1920) included a master plan that suggests little of its eventual complexity and detail. The design proposed a series of roads—some curving, some straight—and a neoclassical chapel by Lewerentz with a nave set perpendicularly to the narrow axis it was to terminate.

Asplund's own architectural contribution from this period, the tiny Woodland Chapel, exemplified an attitude toward building in the landscape characterized by an extreme sensitivity to both place and occasion [3-9]. The massing of the chapel suggests the wooden country churches that dotted southern Sweden in the seventeenth century. Its sophistication, in turn, reflects the polish of the Hytten at the Liselund villa on Møn in Denmark (built from 1792–1795 by Andreas Kirkerup), a structure that received renewed attention in Scandinavian architectural circles early in the twentieth century. The use of wooden shakes as a roofing material—also a traditional practice—determined the steeply pitched roof that dominated the image of the church. Simple columns of a proto-Doric order elegantly supported the shingled roof of Asplund's chapel. More important than the chapel's architecture, however, was the integration of the chapel compound with the remainder of the forest site.

A circumferential wall secures the chapel's courtyard, a wall too high to see over, too solid to see through. And yet, almost in the Japanese manner, one glimpses the roof even beyond the wall. As one enters the gate, the expectation of a controlled world, or a courtyard of omission, is unfulfilled. Asplund has permitted the forest to penetrate and continue through the confines of the court. The architect typically spurned a simple yes or no decision and instead substituted ambiguity and softness by admitting trees into what would normally be a more formally treated space. The wall and the resulting courtyard inflect the texture of the forest, but they do not oppose it.

A crypt has been dug into the earth within the walled area before the chapel [3-10]. Asplund sought no physical conflict with the chapel; the resurrection of the soul of the deceased and hope for the living, not the mortal body, took precedence. Burying the crypt reduced it to a mere inflection of the earth's surface. Encountering the rough, shingled exterior of the chapel one envisions a folksy, dark, brooding, wooden interior in the manner of the country church. Instead, the chapel radiates the pure rationalism of the cube and the hemisphere [3-11]. Light pours down through the rooflight, flooding the dome and the interior beneath it. The eye is drawn magnetically upward. The architect has orchestrated the sequence of movement and view; the path leads from the natural, to the inflected natural, to the highly wrought, to the celestial.

From 1935–1940, Asplund alone completed the transformation of the landscape at Skogskyrkogården and a series of larger chapels. The canonical view of the largest of these, the Chapel of the Holy Cross, is justifiably famous, and the place is imbued with a sense of the religious independent of the iconography of the great stone cross [3-12; see also figure 3-18]. In totality, the cemetery is marked by repose and silence, by the uncanny calm that Ingmar Bergman depicted in his film *The Seventh Seal*, the calm that comes just before the first rays of dawn, just after the first snowfall, or when twilight turns to night.

The site-plan of the complex shows only one line that might be taken as an axis although the straight wall of chapels and burials provides the planar splint that structures the landscape as an entity apart from the random spacing of the pine forest [3-13]. The cemetery's entry area—Lewerentz's work—is uncompromisingly rigid, but, again, in an almost Japanese manner it softens to semiformal and informal order within the precinct of the cemetery proper.

[3-9]
WOODLAND CEMETERY,
SOUTH STOCKHOLM.
WOODLAND CHAPEL.
1918–1922.
ERIK GUNNAR ASPLUND.
The chapel's profile
recalls the country
churches of seventeenth-
century Sweden, but its
execution is more
decidedly classical.

[3-10]
WOODLAND CHAPEL.
The crypt, literally of
the earth, reinforces the
release of resurrection
to come.

[3-11]
WOODLAND CHAPEL.
Within the small chapel,
the pure world of the
hemisphere and the
flood of light invoke a
sense of the celestial
and the pure.

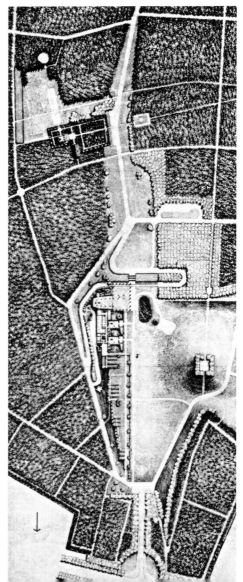

[3-12]
WOODLAND CEMETERY.
THE MEADOW AND
CHAPELS. 1915–1940.
ERIK GUNNAR
ASPLUND AND
SIGURD LEWERENTZ.
The path; the stone
cross stands in the
distance.

[3-13]
WOODLAND CEMETERY.
PARTIAL SITE PLAN.
1940.
ERIK GUNNAR ASPLUND.
The entrance to the
cemetery is at the
bottom of the drawing.

Today one leaves the subway train at the station "Skogskyrkogården." The station and its immediate environment are typically modern; nothing much marks them as special. To the right of the station a small flower stand provides the first hint of the cemetery's presence, and beyond it extends a row of rigorously pleached trees, clipped and fixed in time, that defines the spatial corridor leading to the semicircular exedra at the main entrance. From this point the road leads directly into the cemetery grounds. Two stone walls frame the road and entrance; the iron gates remain fully open during the day. The left wall houses a small classical portico that in turn encases a fountain [3-14]. The fountain neither drips nor flows and the slow movement of its waters is barely sufficient to mark the passage of time—there is something of the eternal about it. Turning from the fountain, looking through the confining stone walls, one is led into the void and to the grass below and the sky above [3-15].

The order feels natural. From the rigidity and the straight geometry of the retaining walls, which like blinders on a horse focus the eyes directly ahead, one is pulled into the soft undulations of the gently rolling dale. In summer the grass is very green and the sky is very blue [3-16]. Powerful cloud formations often pepper the azure dome and give it contour. But our attention falls almost immediately on the stocky, stone cross in the distance; detached from the chapels, it stands as an independent monument.

Quickly now we are drawn to the line of chapels and their stone walls running to our left, paralleling the path—a line across the land that runs over the hills, through the woods, to the horizon. But to our right, atop the knoll some way off, a formation of trees appears too regular to be natural, yet its vegetation seems too unkempt to have felt the hand of the gardener [3-17]. The knoll, too, straddles the edge between the natural and the made.

We proceed to the left along the stone path through which the meadow's grass meanders. Along the way we may glimpse the urn burials within the walls. Our eyes, directed straight ahead, rest in the interval between the cross and the chapel; the view remains unterminated or terminated only by the heavens above. Nearing the building we note the portico, the sculptural group thrusting through the open atrium that provides both the needed illumination and a hint of the impending resurrection [3-18, 3-19]. We move from the open field, to the restricted spatial channel, to the roofed portico, to the open sky above, through

[3-14]
WOODLAND CEMETERY. The slowly dripping fountain built within the ashlar walls.

[3-15]
WOODLAND CEMETERY. The high stone walls direct all vision ahead, into the void of the meadow.

[3-16]
WOODLAND CEMETERY. Emergence into the void: the meditation knoll to the right and the Chapel of the Holy Cross to the left.

the doors to the chapel. The totality is contrived with consideration and architectonic mastery.

From the knoll one can look back to the Chapel of the Holy Cross and view the buildings that are so carefully fitted into the landscape —and yet so strongly expressive of the architectonic order. The grove of birches to the south of the knoll also appears natural at first, although it has been laid out and planted with the mathematical regularity of an orchard. Its configuration echoes, in natural materials, the rhythm and dimensions of the portico colonnade that it addresses across the meadow.

But this is summer.

In winter another world emerges [3-20]. What was green is now white, and the buildings are the darker rather than the lighter forms. Trees that were mass are now linear skeletons. It is beauty of perhaps an even higher order, particularly in the afternoon when the sun slashes across the land and almost parallels its contours. Light defines the soft topography of the surface. This place would feel sacred even to someone lacking an understanding of Christian religious symbols.

At Woodland one senses the coexistence of people and nature as part of a continuity. One order comments upon and, in turn, defines the other—like the figure and ground of a two-dimensional composition. It is difficult, almost impossible, to look at one without looking and sensing the existence of the other. This continuous reflection and alternation of perception, rather than the architectural forms alone, ultimately yield a state of harmony, equilibrium, and aesthetic emotion.

Across the Gulf of Bothnia in Finland, Alvar Aalto began architectural practice in the 1920s, and in his first decade of practice he constructed two landmark modernist buildings: the Turun Sanomat newspaper offices and the Paimio tuberculosis sanatorium. Like the Stockholm Exhibition—a project that influenced Aalto—the sanatorium introduced Finland to full-tilt modernism in a manner and to a degree that would not be equaled until many decades later.

Just as Asplund appears uncomfortable with the severely architectural order within the natural setting, so Aalto shares an equal distaste for the completely rectilinear. As early as the Paimio project he had dis-played some of Le Corbusier's delight with the contrast of the curve against the prisms of the main block. The foil of the free-form entry canopy or the later reception booths play against the straight planes

[3-17]
WOODLAND CEMETERY.
The trees on the knoll read as the natural precedent for the columns of the chapel's portico.

[3-18]
CHAPEL OF THE HOLY CROSS, WOODLAND CEMETERY. 1940.
ERIK GUNNAR ASPLUND.
Seen from the meadow the portico, chapel, and cross reflect things terrestrial.

[3-19]
WOODLAND CEMETERY.
The open atrium of the portico admits light to the sculptural group and allows the eye to rise upward unimpeded.

[3-20]
WOODLAND CEMETERY.
In winter light.

of the walls, as does the sensuous wave of the lecture hall ceiling at the roughly contemporary Viipuri library.

One could find no better example of this particular formal predilection than the "typical" Aalto libraries at Seinäjoki or Rovaniemi. In these buildings the rectangularity of the offices, classrooms, service rooms, and entrance oppose the fan-shaped, phototropic angularity of the principal reading rooms. One can see that, like the made versus the natural, the presence of the straight defines the presence of the curve and vice versa.

In an analogous manner, Aalto used architecture as a condensed and often orthogonal contrast to nature, and the natural as a mitigating influence upon architecture. In his studio in Munkkiniemi, for example,

the interior court can be read as a great bite carved from the quiet rectangular block that fronts the street. Within this concavity, the earth is terraced in stepped contours to fashion an architecture of earth. A more dramatic application appears in the great grass staircase at Säynätsalo, where the steps actually geometricize the natural slope and those contours resulting from filling the central court [3-21]. Here, the earth is permitted to ooze out from the upper level in a controlled and rationalized manner. At Seinäjoki the earth mounds against the south wall of the town hall in sweeping angular contours that reappear, in profile, as the roof of the council chamber. The fluid curve of nature becomes, in Aalto's hand, the segmented curve, more easily constructed, rationally determined, an appropriate transition between nature and architecture.

[3-21]
TOWN HALL,
SÄYNÄTSALO, FINLAND.
1952. ALVAR AALTO.
Rationalized contours
shape the grass stairs,
which reflect their
granite counterpart on
the opposite side of the
courtyard.

The Villa Mairea, built in 1939 near Noormarkku in western Finland, illustrates many properties of Aalto's involvement with the curve and the inherent order in nature. The villa was constructed for Maire and Harry Gullichsen, noted patrons of the arts and the inheritors and builders of a considerable industrial conglomerate. As a family estate the site already possessed two major residences that included a sprawling villa built about the turn of the century. Pine woods cover the land; grass grows underfoot. The topography exhibits few dramatic changes as the land slopes gently over its bed of granite.

A nearly regular grid of columns supports the interior spaces of the house and augments the structural capacity of the masonry walls. The

treatment of the columns, however, betrays little regularity. Within the villa Aalto has essentially rationalized the forest; from the very first to the very last, no column has been left untreated: modifications link one column to another or distinguish it from the next. Some columns are primarily structural, yet are treated aesthetically; others are used only for architectonic purposes.

Approaching from the south we view the principal facade with its angled bay windows and the free curve of the projecting entrance canopy that recalls the entrance to the Paimio sanatorium [3-22]. Upon closer investigation we note that the canopy is curiously supported: On the right a thick concrete column no doubt provides the primary support. But the column has been wrapped with unpeeled saplings. To the left

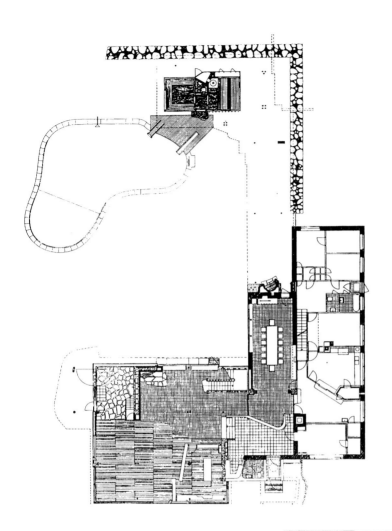

[3-23]
VILLA MAIREA. PLAN.
The regular spacing of
the columns structures
the individual, unequal
treatment.

[3-24]
VILLA MAIREA.
A screen of poles at the
lower entrance level—
stripped of bark in this
variation—leads to the
living room beyond.

[3-25]
VILLA MAIREA.
The twin black-lacquered
columns in the "winter"
area of the living space
are bound in rattan to
the height of the fire-
place mantel.

[3-26]
VILLA MAIREA.
In the "spring" area of
the living room the
columns vary in cluster-
ing and wrapping.

[3-27]
VILLA MAIREA.
The vines covering the
principal stair complete
the metaphor of the
column and the forest.

poles of slightly larger diameter have been joined by vine lashings, strengthening the sense of structure. One recalls the more romantic aspects of Aalto's work such as his pavilion at the 1937 Paris Exposition and the earlier Turku Fair, and also to the stylistic period 1880–1905 known in Finland as National Romanticism, during which various architects and artists built their own villas after the manner of vernacular farmhouses.

Entering the house at the lower level, the view is carefully directed [3-23, 3-24]. A low wall defines the entry, while a field of non-structural poles comprises a screen that recalls the saplings encountered just outside the front door. Some poles fall on the entry side of the wall and some on the opposite—Aalto obscures the clarity of the intervening wall, and the spaces begin to blend. Mounting the few steps toward the living room the eyes rest on a pair of lacquered columns standing as a pivot to the space [3-25]. They are wrapped with split rattan, a treatment that Aalto would use in many later projects, particularly in civic buildings when he substituted marble or granite for the fragile rattan. Deeper in the space other double columns are bound in tandem [3-26]. Throughout the lower floor no two columns appear alike although they are regularly placed and share the same palette of materials.

Turning back, one encounters the main staircase to the second floor inserted between the living and dining rooms. It is contained within its own forest of poles, some of which are plugged together like the canopy saplings at the entrance. Vines have overgrown the staircase, completing the arboreal analogy [3-27]. The structural approach is augmented by a myriad other considerate details used throughout the site, such as the fence and hillside that articulate the pool area as a constructed mound, or the sod roof on the sauna and connecting loggia.

Reminiscent of the path from the forest to the light at Asplund's early Woodland Chapel, Aalto brings the natural order into the house, reforms it, and rationalizes it through a gradual and consistent restructuring of its form. The transition is smooth. The house is articulated not only as a place with its own order and sense of identity but also as a place that shares certain morphological sympathies with the forest around it [3-28].

At the Woodland Cemetery and at the Villa Mairea construction stands apart from the natural order and yet fits comfortably within it. Architecture does not assume the guise of an isolated machine in the garden; instead building and landscape demonstrate a synthesis of respective orders, elements, and qualities. The edge between the two is purposefully ambiguous and blurred. As one mentally squints at the pieces certain components of the setting snap sharply into focus while other parts soften and become cloudy. In the end, we cannot quite rest in a state of complete knowledge nor can we accept as exhaustive and unchanging the information and sensations provided. Neither can we accept as given and finite the reactions and emotions to which those sensations are linked.

Originally published in *Places*, Volume 1, Number 2, 1984.

[3-28]
VILLA MAIREA.
The resonance of house
and forest.

Acknowledgments

I wish to thank the following people for their help in the preparation of this article: for critical review, Professors Lars Lerup and Mary McLeod; for sharing his knowledge of Sigurd Lewerentz's work at the Woodland Cemetery, Professor Janne Ahlin; and for photographs, Kristiina Nivari and the Museum of Finnish Architecture, Helsinki.

Travel and research for this study were aided by sabbatical funding from the University of California, Berkeley, and a Senior Fulbright Lectureship from the United States Educational Foundation in Finland, 1982.

4

The Presence of Absence: Places by Extraction

1987

Every act of design constitutes an act of disturbance. Whether the pre-existing field is natural or contrived, construction must necessarily disturb that which has been. The term *disturbance* should not be regarded negatively, however, since in this broader sense disturbance can actually improve the existing state or provide an equally appropriate —or even improved—alternative. Not always, of course, but sometimes.

Psychology tells us that we first perceive by contrast: we read the outline of the tree before we discern the color and shape of its leaves; we note the presence of the sound before we ascertain the rise and fall of its melody. Perception is at least a twofold process: the first stage is an awareness; the second attends more directly to the particular stimulus.[1] In staking out our place in the world, we begin by reforming the prevalent order as a means of overlaying significance upon that terrain.[2] Disturbance, whether by adding or subtracting from the landscape, is the first consequence of environmental design.

In the developed world, we tend to regard construction primarily as an act of displacement. The mass of a building occupies real space in the wilderness by replacing rocks, trees, or earth; in the city, buildings replace open space or other buildings. But there is an alternate way of looking at making architecture and places. Instead of adding more elements to the site, let us consider rearranging that material already present or removing material from it. Perhaps by looking obliquely at the more common process of addition, we can more readily accept an existing condition as worthy of continuance or build for an economically greater return.[3]

The power of absence does not depend on a conscious aesthetic intention. […] In the forest, the dense, inarticulate texture of the trees provides the basic ground against which design intention must be measured. In Japan, forests blanket the vast majority of the land surface, the valleys and the plains but especially the mountain slopes. During the early centuries of the Christian era, the hegemony of the Yamato clan consolidated on the land around what today is Nara and Kyoto. The gradual separation of the practice of Shinto religion from the person of the reigning chieftain—in time, the emperor—necessitated religious structures distinct from residential types. Although the architectural style of the high shrines derived from archaic storehouse forms, the disposition of the shrine precincts drew upon principles of geometric order foreign to the indigenous, topographically derived, precedents.[4]

[4-1]
NAIKU (INNER SHRINE).
ISE, JAPAN.
FIFTH CENTURY;
1973 REBUILDING.
Although the wooden
structures and concentric
fences overlay the sanctified
space, removing trees
constitutes the primary
act of place-making.

Shinto deities dwell in places at times announced by unusual geographic features. Shrine architecture taps upon and articulates the power of the site rather than creating an internalized sanctuary representative of Christian tradition. At the Shinto shrine all visitors pass through the sanctified precinct, but the most sacred celebrants alone are privy to the interior of the centermost structures. To differentiate the zone of the shrine, the order of the site is consciously disturbed, a practice common to many peoples throughout the world. "When the sacred manifests itself into any hierophany," explained Mircea Eliade, "there is not only a break in the homogeneity of space; there is also revelation of an absolute reality, opposed to the nonreality of the vast surrounding expanse. The manifestation of the sacred ontologically founds the world."[5]

[4-2]
GEKU (OUTER SHRINE). ISE, JAPAN. ALTERNATE SITE. 1973 REBUILDING. Lining the floor with white gravel distinguishes the rectangle of the zone from its forested surroundings. Note that certain trees remain within the purity of the geometric figure.

At the Ise shrines, since the fifth century continually reconstructed in almost the exact form every twenty years, the first act is to clear and demarcate a zone within the forest [4-1]. The pure geometry of the rectangular zone—divided into two sites and built upon in alternating cycles—is softened by the native Japanese acceptance of natural incident. Some trees remain within the cleared field or on its perimeter [4-2]. Geometry is not taken as a pure abstraction but becomes instead an abstract ordering tempered by the particularities of the site's topography and the cryptomeria forest. Carpeting the rectangular zone, white gravel reinforces the sense of limits by vividly contrasting in texture and brilliance with the soft earthen and needle-covered forest floor.

[4-3]
RYÔAN-JI.
KYOTO, JAPAN.
CIRCA 1499, WITH
LATER CHANGES
The garden clearly reads
as a void against its
vegetative and topographic
backdrop.

From this point, the operations become more architectural, adding in concentric layers an arrangement of four fences of varying density, and building the wooden shrine structures themselves. In many ways these buildings are secondary to the primary definition of the sites at Ise, however. More than any other single act, the removal of the preexisting forest—disturbance—heightens the presence of the sacred space.

The Shinto tradition of suggestion has continued into the modern era. One can conceive of absence as omission or one can think of it as abstraction. Normally we consider dry gardens such as the famous rock garden of Ryôan-ji (circa 1499, with later modifications) in Kyoto as absence: only gravel, moss, and fifteen monumental stones comprise the essence of this landscape [4-3]. In looking at the sheet of gravel

and the subjects embedded in its surface, we tend to think of it two-dimensionally, that is, as a plane. One must, however, consider the garden in context. Seen against the surrounding earthen wall, the flowering cherry trees just beyond the garden, the maples that assume brilliant colors in autumn, and most of all the adjacent hillsides, the rock garden is less a single plane and more a spatial void, as Ise occupies the void in the forest. In Zen, the tradition of the *yuniwa*, the gravel field, thrived beyond the confines of Shinto and has become a basic element of the Japanese garden vocabulary.[6]

The act of removal at Ise functions as a perceptual trigger. At other sites the landforms recall the trace of prior use; like the forest cut in the Tatras, they are not necessarily aesthetically intentioned. The peat bogs on the Shetland Islands, for example, assume the shape of rectangular lots: the peat is incised, extracted, and burned in rectangular blocks [4-4]. The harvesting of the peat—with each slab assuming a relatively constant dimension in thickness, depth, and width—over time creates a negative pit of orthogonal geometry, a continually varying relief that unwittingly maps the diminishing natural resource. In the summer months, with the need for heating fuel greatly reduced, the grass triumphs—integrating the myriad cuts into an abstract relief of blocky, irregularly stepped contour, both soft and hard at the same time.

[4-4]
PEAT BOGS.
SHETLAND ISLANDS,
GREAT BRITAIN. 1969.
In time grass covers
the site, transforming
the excavations into a
rectilinear relief, the
record of human activity.

Here sight parallels sound. At times it is silence rather than noise that exerts the greater presence. Indeed, the Japanese say that a whisper can be heard when a shout cannot. Composer John Cage once wrote

that "the music never stops, we just stop listening." Thus, it is natural, rather than completely ironic, that the composer entitled a book of his writings *Silence*. Cage, who has been influenced by Asian religion and aesthetics as well as Western mysticism, also created a piece entitled *4' 3"*, "a silent piece in three movements."[7] Of course, there is no true silence; Cage tells us that we will always hear something, if nothing other than the normal surroundings of our lives that ordinarily are left unheard, or the pulsing of the blood in our bodies. Absence is presence in his music: "We are in the presence not of a work of art which is a thing but of an action which is implicitly nothing," Cage wrote about the music of Morton Feldman. "Nothing has been said. Nothing is communicated. And there is no use of symbols or intellectual references. No thing in life requires a symbol since it is clearly what it

[4-5]
DOUBLE NEGATIVE.
MORMON MESA,
NEVADA. 1967–1970.
MICHAEL HEIZER.
The sculpture is created
solely by excavation and
displacement, with two
aligned cuts into the mesa
edge visually linked.

is: a visible manifestation of an invisible nothing."[8] Similarly, there are no visual silences in the environment. By simplifying, by abstracting, we may focus but we may never reach point zero.

During the 1960s a number of artists left the confines of the art gallery to create works in open space. To some of these artists, creating works that could not be marketed was a political statement against treating the art object as yet another commodity of capitalist society. But questions of scale and content also occupied central positions in the conception of these works. Robert Smithson directed us to re-examine the quotidian environment—often taken as ugly—to expose a reality fraught with considerable power and aesthetic possibilities.[9] Sculptors such as Michael Heizer have depicted the order of a situation through

the structure of their works. Heizer's series of excavated displacements, for example, removed portions of the desert floor to reform the surface into a noticeable configuration while unveiling the geologic composition below the surface. For Heizer, "the subject matter of sculpture is the object itself, sculpture is the study of objects." But, "a statement about anything physical becomes a statement about its presence." The "drawings" (*Nine Nevada Depressions*, late 1960s) engraved in the desert floor are "sculptures with weights removed."[10]

Double Negative (1967–1970) is the largest of Heizer's works in the Nevada desert to date: two mammoth straight channels in a mesa align across the irregularly eroded edge of a cliff [4-5]. To make the cuts, bulldozers scraped two sloped ramps, each incision pushing deeper below the Earth's rocky surface, dumping the excavated material over the edge in a manner paralleling Smithson's *Asphalt Rundown* (Rome, 1969).[11] The residue of Heizer's process is a broken spatial channel approached by ramps at either end. As one descends the ramp, the desert disappears and in its place, like a pair of gargantuan blinders, the sides of the cut rock rise into one's consciousness. The unreinforced sides of the channels expose the sedimentary strata and reveal the effect of time on the development of the land. In the descent, one is, in effect, traveling backward in geologic time, while one's view is focused across the opening toward the reciprocal void. In this work, the sculptor has done little more than disturb the condition of the mesa's edge; the straight cuts betray the human presence as the strata elucidate the geology.

Richard Serra's *Casting* (1969) was created by slinging molten lead against the intersection of a wall and floor. As the metal cooled, the sculptor pulled the linear casts from the intersection which had become its form. The piece's configuration—the straight edge and the rough edge, the irregular grain produced by the successive throws—lucidly recorded the process of its making. Serra regarded the room as the negative, using the mold for his positive casting.[12] Heizer, on the other hand, considered the desert as a positive from which he removed two linear sections to create a the desired void. It is just that absence of the rock in *Double Negative* that creates its vitality [4-6].

While we read certain urban open spaces as positives, we read others as voids. John Wood the Elder's 1736 Queen's Square or John Palmer's

[4-6]
DOUBLE NEGATIVE.
The channels remain evident after almost thirty years of erosion.

[4-7]
LANSDOWN CRESCENT.
BATH, GREAT BRITAIN.
1789–1793.
JOHN PALMER.
The building wraps the air to enclose space, but one senses that the architectural definition is thin.

1793 Lansdown Crescent in Bath wrap new architecture around air to form places [4-7]. Although one may feel that they are only as deep as the buildings themselves—open places girdled by buildings—the spaces and the architecture that define them may yet read as positives.

In contrast, there are urban spaces that thunder as voids: with the noise of silence and the power of absence. The populace must have read the drastic cuts of Haussmann's Parisian renovations or the *sventramenti* of Mussolini's Rome in just that way: brutal incisions on the body of the city whose new order defied the existing texture like the great high-voltage easements that cut mercilessly through the forests. Mussolini, an impressive and powerful orator, also understood the manipulation of spatial context. His redevelopment projects often called for the isolation of ancient monuments in their "necessary solitude" to heighten both their presence and the role of the present in the sweep of Roman history.[13]

Of all the world's urban spaces perhaps none reads more powerfully than the justifiably oft-mentioned "drawing room of Europe:" the Piazza San Marco. Populating the Venetian lagoon was a trying task, and dwellings, churches, and other structures came to occupy every available square meter of dry land. Only the *campo,* around which a neighborhood centered, and narrow paths remained between the densely packed structures. While there are many *campi*, there is only one truly great piazza. Fronting the basilica and linked perpendicularly to its smaller piazzetta, the Piazza San Marco appears as a vacant site awaiting to be filled. One feels the tension, the long *procuratie* wings acting as retaining walls against the thrust of urban development [4-8]. Like the police cordon, they hold the city at bay, allowing the crowds to gather on the piazza, to traverse its length and breadth, to promenade, to view the church, deal, take coffee, or watch the pigeons and tourists. In a manner that recalls the extraction from the forest at Ise, the Piazza San Marco can be sensed as a removal from its urban fabric, in feeling if not in historical fact.

The power of absence is felt in varying contexts. We notice for the first time certain structures when they have been torn down for a parking lot or when the site is vacant, awaiting construction. Like the exaggerated sense of the missing tooth, one becomes more aware of it after its extraction. Departure from the normal order, whether it be construction in the natural setting or destruction in the urban environment, controls our attention. We can focus only on the void, at times forgetting what has been removed.

One artist concerned with calculated removal was Gordon Matta-Clark (1943–1978). Matta-Clark's late work provided conceptual transparency in an opaque environment. Using a chain saw to cut through structures —usually buildings marked for demolition—Matta-Clark simultaneously revealed the building physiognomy while formulating new spaces comprised of the voids. Perhaps his most powerful work was *Circus* (1978, also called *The Caribbean Orange*), created by cutting and removing pieces of a structure adjacent to the Museum of Contemporary Art in Chicago slated for renovation [4-9, 4-10]. By cutting through the floors and walls in a series of varying shapes based on the arc, the composite space emerged, distorting and billowing as it traversed the derelict structure's three floors.

His was a conjuring trick, making three-dimensional spaces using two-dimensional layers of space. Judith Kirshner wrote the following about Matta-Clark's sculpture:

> Like spiral forms, the dynamic volumes Matta-Clark carved in these last major works gave the feeling of being endless. They eluded comprehension as one passed haltingly through the spaces, climbing up and down, walking to and fro, even jumping as one looked. On the third floor of *Circus*, a truncated section of a sphere, a circle of Sheetrock with a door in the center, was dramatically suspended as if to defy gravity, architectural reason, and visual understanding. Matta-Clark often spoke of the apprehension of his multilayered works being dependent on recollection, of the impossibility of their being instantly assimilated.

[4-8]
PIAZZA SAN MARCO.
VENICE, ITALY.
One feels the tension along the plaza's outer edge, which like a wall retains the pressure of urban development around the square.

[

[4-9]
CIRCUS (OR *THE CARIBBEAN ORANGE*).
CHICAGO. 1978.
GORDON MATTA-CLARK.
The spatial figures and sculptural definition of the work derive from selectively removing portions of a structure slated for renovation.

[4-10]
CIRCUS (OR *THE CARIBBEAN ORANGE*).

This, however, was no cause for despair: "of course, it recognized that fragments can be more telling than totalities."[14]

We have all seen urban wall remnants upon which the records of the once-abutting spaces have been deposited on the party walls of their neighbors. The incongruence of residual ceramic tile, the curiously tinted plaster surfaces floating on brick, or the fragments of residual concrete overlay scale and history to the normally blank walls that turn away from the street. Matta-Clark's work provides us a similar chronicle of building history, augmenting residue with a vision of the positive void that charges through floor planes, claiming space and identity. Incisions reactivate memory, proving that the power of the void can supersede architecture's repository of pragmatics.[15]

The act of construction is an act of covering; each addition overlays an existing condition and grants it a new configuration. Enclosure usually disguises structure, a favorite target of modernism's call for truth in building. Memory, too, plays a role in creating the presence of absence, for we must know or remember what has been before we can fully comprehend what is now. Perhaps this need for recall engages us in a more active discourse than those acts of addition normal to architectural construction. The void induces us to participate in ways that the solid cannot. Intrigued, we question just what is going on here, just what has changed, just what is different in the picture. Party walls exposed after building demolition tell us—in section—the story of the building now passed, rendered transparent for the first time since it was enclosed by the act of construction. Demolition can serve as an act of revelation [4-11]. Gordon Matta-Clark and Michael Heizer, in their sculptures, clarify by "dis-covery" and provide us with a lens with which to see, as if enlarged, the world we usually pass without notice.

Originally published in *Places*, Volume 4, Number 3, 1987.

Notes

1 See Ulrich Neisser, *Cognition and Reality: Principles and Implications of Cognitive Psychology*, San Francisco: W. H. Freeman, 1976.

2 Marc Treib, "Traces Upon the Land: The Formalist Landscape," *Architectural Association Quarterly*, Volume 11, Number 4, 1979 [included in this volume].

3 Economy here should be distinguished from cheapness. Cheap refers to the lowest possible cost in spite of return; economy is the greatest return for the amount of resources invested.

4 See Günter Nitschke, "Ma: The Japanese Sense of Place," *Architectural Design*, May 1966.

5 Mircea Eliade, *The Sacred and the Profane: The Nature of Religion*, New York: Harcourt, Brace & World, 1959, p. 21.

6 See Teiji Itoh, *The Japanese Garden*, New Haven, CT: Yale University Press, 1972, pp. 142–147; and Marc Treib and Ron Herman, *A Guide to the Gardens of Kyoto*, revised edition, Tokyo: Kodansha International, 2003, pp. 4–6.

7 See Anne d'Harnoncourt, *John Cage: Scores and Prints*, New York: Whitney Museum of American Art, 1982, unpaginated.

8 John Cage, "Lecture on Something" (1959), in *Silence*, Cambridge, Mass.: MIT Press, 1966, p. 136.

9 Robert Smithson, "A Tour of the Monuments of Passaic, New Jersey" (1967), in *The Writings of Robert Smithson*, New York: New York University Press, 1979, pp. 52–57.

10 Michael Heizer, Interview, in Julia Brown, ed., *Michael Heizer: Sculpture in Reverse*, Los Angeles: Museum of Contemporary Art, 1984, pp. 30–31.

11 *The Writings of Robert Smithson*, pp. 192–193. Both works involve the residue of spillage and therefore are physically related, but more importantly, both concern the process of entropy and its effect on sculpture.

12 See Clara Weyergraf, *Richard Serra: Interviews, etc., 1970–1980*, Yonkers, NY: The Hudson River Museum, 1980; and Rosalind Krauss, *Richard Serra/Sculpture*, New York: The Museum of Modern Art, 1986.

13 Spiro Kostof, *The Third Rome: 1870–1950: Traffic and Glory*, Berkeley, CA: University Art Museum, 1973.

14 Judith Russi Kirshner, "Nonuments," *Artforum*, October 1985, p. 103. See also Mary Jane Jacob, *Gordon Matta-Clark: A Retrospective*, Chicago: Museum of Contemporary Art, 1985, especially pp. 112–130.

15 *More recently, Rachel Whiteread has explored the casting of negative spaces—in plaster, concrete, or resin—transforming the void into solids, the negatives into positives. See Fiona Bradley, ed., Rachel Whiteread: Shedding Life, Liverpool: Tate Gallery, 1997; and Rachel Whiteread: Transient Spaces, New York: Solomon Guggenheim Foundation, 2002.*

[4-11]
DEMOLISHED BUILDING
SITE.
BERKELEY,
CALIFORNIA. 1987.

[5]
Formal Problems

1998

English-speaking writers on landscape design and its histories are dogged by a particularly nasty linguistic problem: how to define and employ the word *formal*.[1] Unfortunately—alas—its definitions occupy two very different arenas. In one zone, we find formal defined in relation to shape or design; as the dictionary puts it: "relating to or involving the outward form, structure, relationships of, or arrangement of elements rather than content." "Style of painting" is offered as an example.[2] On the other hand, formal can be posited as the antithesis of "informal," that is, "characterized by a punctilious respect for form" [5-1]. This second range of definitions, however, is almost tautological: formal as a punctilious respect for form. Despite the ambiguities that haunt both applications of "formal," using the word tends to be unavoidable in the study of the designed landscape, compounding enormously our perceptions and critical thinking.

Now, neither of these definitions of formal need apply to a specific manner—or *style*, if you will—of landscape design. And yet to a large degree this posed antithesis of formal versus informal manners has directed our construction of the history of landscape architecture, landscape design, or landscape gardening (call it what you will). Formal design, as we know from readings on André le Nôtre if from nothing else, involves the straight line, usually geometry, and often symmetry [5-2]. It assumes an architectonic stance, and it anthropomorphically asserts the human mind and hand. Informal design, it is said, instead attempts to underplay or even conceal all signs of human contrivance [5-3]. Here nature remains the ultimate model for the landscape's design, although this root concept may not always be apparent. Composition emulates the natural order; plants, we assume (however fallaciously), are allowed to grow as God intended them, and the net effect is that of a natural setting without the artifice of human intervention.[3] While only God can make a tree, certainly a human being can make a scene in which that tree appears to remain in its natal matrix. Is this not a very formal intention?[4]

Certainly, all of these are to some degree naïve assumptions, and establishing such oppositions suggests post-Enlightenment, bipolar thinking more than any practice inherent to the making of landscapes. Rarely, if ever, do we encounter the presence of the formal independently of the informal. Rarely, if ever, can we truly create a natural setting, so "informal" that it would be mistaken for nature itself. Do these oppositions of formal and informal (which can be translated, at the very least, into the natural and the constructed) really exist? They do;

or at least to some degree. But they have tended to be more important in discourse than in actuality.[5] To trace some of this ebb and flow of formalities, let us review several celebrated sources, beginning with two from late eighteenth-century France.

The scene is René-Louis Girardin's own garden at Ermenonville, created by the marquis himself and fueled by the writings of Jean-Jacques Rousseau, who sought our origins in the nature of a natural garden [5-4]. In his oft-cited *Julie, ou la Nouvelle Héloïse* of 1761, Rousseau argued that as a natural entity the garden must seek an affinity with nature. In the chapter "Letter to Lord Bomston," the narrator relates his having entered a garden only a few footsteps from the house; yet is it another place, the far more charming world of an orchard transformed: "You have locked the door, water has come I know not how, nature alone has done all the rest, and you yourself would never be able to do as well." Later on, the nearly overwhelmed guest exudes that "You see nothing laid out in a line, nothing made level. The carpenter's line never entered this place. Nature plants nothing by the line. The simulated irregularities of the winding paths are artfully managed in order to prolong the walk, hide the edges of the island, and enlarge its apparent size, without creating inconvenient and excessively frequent turnings."[6]

Girardin listened carefully to the philosopher whom he hoped would find eternal rest at Ermenonville. In his 1777 treatise, *De la Composition des paysages*, Girardin drew in equal measure from Rousseau's philosophy and his own practical gardening experience. In the text (that is, in the 1783 translation by Daniel Malthus), Girardin first distances himself from the achievements of the previous century's *jardin à la française* [5-5]:

> The famous Le Nôtre, who lived in the last age, contributed to the destruction of nature by subjecting everything to the compass; the only ingenuity required, was measuring with a ruler, and drawing lines like the cross-bars of a window; then followed the plantation according to the rules of symmetry; the ground was laid smooth at a great expense, the trees were mutilated and tortured in all ways, water shut up within four walls, the view confined by massy hedges, and the prospect from the house limited to a flat parterre, cut out into squares like a chess-board, where the glittering sand and gravel of all colors, only dazzled and fatigued the eyes; so that the nearest way to get out of this dull scene, soon became the most frequented path.[7]

[5-5]
VERSAILLES. FRANCE.
LATE SEVENTEENTH
CENTURY.
ANDRÉ LE NÔTRE.
The principal axis.

Here formality is paired with the ruthless destruction of the woods and countryside—themselves the beguiling antidotes to the contrived geometry of the parterre and *pièce d'eau.*

To the marquis the answer to this formal effrontery is nature herself. "Wherever this taste is introduced," Girardin offers, "nature will smile with all the graces of elegant simplicity, its infinite variety will never cease to amuse, and it will produce the secret charm of which no feeling mind can tire." Symmetry has no place in this garden, its having derived from "vanity and indolence." The secret? "It is not then as an architect or a gardener, but as a poet or painter, that a landscape must be composed, so as at once to please the understanding and the eye."[8]

We need recall that Girardin was reacting against the formality of, and the autocracy that lay behind, the *jardin régulier*, that is, the formal garden of the time of Louis XIV, give or take a Louis or two. The naturalistic garden manner was thus a conscious political and philosophical assertion, manifested not only in the circuit of the garden's path, but also in the garden's *fabriques*: those structures or follies that punctuated the landscape.[9] At Ermenonville, one encountered a dolmen or prehistoric grotto, a rustic kiosk, the Isle of Poplars with Rousseau's tomb (actually a cenotaph)—with a quite formal planting of trees— and the Temple of Philosophy—in ruins or unconstructed, as you will [5-6]. As architecture, they embodied a gradient of formality ranging from the mere stacking of stones (the dolmen) to true architecture (the temple), and as such, suggested the course of civilization. The form of one structure set the other in contrast, as architecture itself sets in contrast the seemingly natural landscape.

[5-6]
ERMENONVILLE.
FRANCE.
LATE EIGHTEENTH
CENTURY.
RENÉ-LOUIS
GIRARDIN.
Temple of Philosophy.

Girardin's work parallels the fruition of the English landscape garden across the Channel some decades previously. There, too, one was met by sweeping lawns, clustered groves, concealed terminations to bodies of water, and attempts to efface the perceived boundaries of the domain. There, too, were symbolic structures like Stowe's Temple of British Worthies; the Temple of Ancient Virtue intact, but the Temple of Modern Virtue quite in ruins. These formal (i.e. architectural) elements articulated the informality of the lawn and grove. Walpole's noted aphorism that William Kent leapt the fence and saw that all of nature was a garden seems just a bit too simplistic a summation, however, given all this evidence to the contrary. Or, it underscores that Walpole's idea of the garden itself required statuary and, perhaps, building. After all, Kent didn't leave nature alone, he did a lot of reshaping and composing, and he peppered lawns and dales with statuary and buildings. And one can hardly call his 1720s work at Chiswick House naturalistic, at least not by today's standards [5-7]. It did, quite typically, mix artifice and vegetation, that is, formal and informal elements in creating a synthetic work.[10]

Many of the great English estates such as Stowe or Rousham or Petworth all incorporated heroic swaths of grass before the great houses. The more naturalistic of the gardens' zones bracketed these more formally-composed areas, centrifugally spinning ever more informally, blending with the agricultural countryside off in the distance beyond the discernment of optical parallax. Indicatively, two words were applied to these areas: the *garden* (often more decidedly shaped to a plan) and the *park* (at greater scale, more a natural landscape than a designed one). Longleat, a landscape made by Capability Brown from 1757 on stands as a prime example, although its formality was given a booster shot by Russell Page in the 1950s [5-8]. Note how the visually apparent control dissipates as the gaze moves from the building to the pasture: formal enclosures around the house, clipped hedges leading the eye beyond to Brown's detested "clumps," on over the hillsides to the woods and the skies.[11] But even this diminution of legible intervention troubled some.

Enter William Robinson a century or so later, seeking a more natural evocation of the true English landscape, which was to say, the English soul.[12] In *The Wild Garden* of 1870 Robinson implied that the true English landscape was the true English landscape—although he did allow for the introduction of some few exotics.[13] Human management should be restrained, and in its place the beauty of nature should emerge [5-9]. Surprisingly, however, many of Robinson's arguments for the wild garden were quite pragmatic, and are suggested by the book's subtitle. Beauty, he implied, evolved from horticultural logic and design as well as inherent properties: "Hundreds of the finest hardy flowers will thrive much better in rough places than ever they did in the old-fashioned border." On a more religious note he tells us that "there can be few more agreeable phases of communion with Nature than naturalizing the natives of countries in which we are infinitely more interested than those in which greenhouse or stove plants are native."[14] In addition, flowers just look better in a natural context, look better against the green of grasses than the brown of the flower bed with its precise edge. In his writings, Robinson provides precious few hints at the overall order of this natural "work of art;" the garden is more a collection of these horticultural decisions and interventions than a place following the directives of a comprehensive plan.[15]

I would say that the ultimate distinction in *The Wild Garden* was less between formal and informal manners than between gardens and gardening, that is the noun versus the verb.[16] Robinson describes a dynamic process that is experienced in the resulting product; the plan itself is of little importance because a plan is static and represents only a specific moment, usually the moment of hoped-for fruition. But there is no single "garden;" it can exist only in a state of perpetual change. Process outweighs product. To Robinson, the form of the resulting garden is good because the process is just, aesthetics to some degree follow process, and of course, morality. Our experience of the garden overrides any of its elements taken in isolation at any time.

[5-7]
CHISWICK HOUSE.
ENGLAND, 1720s+.
WILLIAM KENT, LORD
BURLINGTON ET AL.

[5-8]
LONGLEAT. ENGLAND.
EIGHTEENTH CENTURY.
CAPABILITY BROWN;
RUSSELL PAGE, 1950s.

[5-9]
"COLONIES OF POET'S
NARCISSUS AND
BROAD-LEAVED
SAXIFRAGE."
ALFRED PARSONS.
[from William Robinson,
The Wild Garden, 1870]

Elements of the Robinsonian argument informed the work of Gertrude Jekyll. But trained as an artist rather than a gardener, Miss Jekyll had no real bones to pick with the ideas of her predecessors. In her introduction to *Wood and Garden* (1899) she allowed that some gardeners may see the site only as a locus for a plant collection. "Others," she wrote, "may like best wide lawns with large trees, or wild gardening, or a quite formal garden, with trim hedge and walk, and terrace, and brilliant parterre, or a combination of several ways of gardening. And all are right and reasonable and enjoyable to their owners, and in some way or degree helpful to others."[17] There is little with which one can take issue in a statement so broadly framed, and general assertions such as these left much room for a personal approach particular to the place.

Perhaps Jekyll's background as a painter led more directly to her involvement with garden color, normally regarded, with texture, as her principal contribution to the making of gardens. She decried the monochromy of the Victorian parterre and instead argued for a play of colors achieved by mixing species. Like the Impressionists, with whom she was contemporary, Jekyll focused on the effect rather than the means—although she possessed a true mastery of her medium.[18] But her interest concerned more than the visual sense alone. The best time to experience a garden, she wrote, was "as the light begins to fall. The early evening hour is indeed the most fulfilling of all; the hour of loveliest sight, of sweetest scent, of best earthly rest and fullest refreshment of the body."[19] Nothing about formal and informal here —any order is appropriate if it leads to a pleasurable consequence (although the phrase "wild garden" crops up frequently in her writings).

In describing her own garden at Munstead Wood, Jekyll implied that the effect of the transitions were paramount: "The intention of all the paths from garden to wood is to lead by an imperceptible gradation from one to the other by the simplest means that may be devised, showing on the way the beauty of some one or two good kinds of plant and placing them so that they look happy and at home."[20] Working in collaboration with architects the likes of Edwin Lutyens, Jekyll used plantings to soften and animate relatively rigid architectonic structures. Her plantings created an interlocked symbiosis of formal—primarily monochromatic and geometric—constructions with those that grow, flower, possess fragrance and constantly change. To the formal matrix, the plantings became an informal counterpoint [5-10].

The Jekyll/Lutyens collaboration at Hestercombe from 1904 to 1909 provides a pertinent illustration of this interplay between the partners and their respective contributions. The architect's terrace and enclosure of stone, symmetrically disposed, used a handsome pergola to engage the adjacent pasture and its bovine habituées. Jekyll occupied this frame with her painting in plants, effecting an augmented coherence among the garden's parts. By disposition, Miss Jekyll was an inclusionist rather than an exclusionist, allowing for a variety of manners within a single system of values. Her impressive output of books on gardening, unlike Robinson's, centered on practical instruction, suggestive and evocative rather than polemical. The polemics, more often than not, were written not by gardeners but by architects, who as we all know, have far more time to write.

For the rhetorical response to Robinson's wild garden, then, we need turn to the architect Reginald Blomfield's *The Formal Garden in England*, co-authored with Inigo Thomas and published in 1895.[21] Blomfield may appear a toady and a prig, but he was far from unintelligent, and his text remains the most cogent and persuasive argument for the reinstatement of the formal garden tradition [5-11]. "The formal treatment of gardens," Blomfield asserted, "ought to be called the architectural treatment of gardens for it consists in the extension of the principles of design which govern the house to the grounds which surround it." In fact, the intention of this approach is to bring the two into harmony. "The landscape gardener, on the other hand, turns his back upon architecture at the earliest opportunity and devotes his entire energy to making the garden suggest natural scenery, to giving a false impression as to its size by sedulously concealing all boundary lines, and to modifying the scenery beyond the garden itself, by planting or cutting down trees, as may be necessary to what he calls his picture."[22] Making an informal garden is really no challenge, Blomfield quips: all you have to do is get the gardener drunk, let him loose in the garden, and then trace his path.[23]

In developing his argument, Blomfield stressed that the word garden itself "means an enclosed space, a garth or yard surrounded by walls, as opposed to unenclosed fields and woods."[24] Blomfield distinguished the horticulturalist from the gardener and the garden designer: "they should work under control, and they stand in the same relation to the designer as the artist's colorman does to the painter, or perhaps it would be fairer to say, as the builder and his workmen stand to the architect."[25] In addition, and perhaps ultimately more important, the

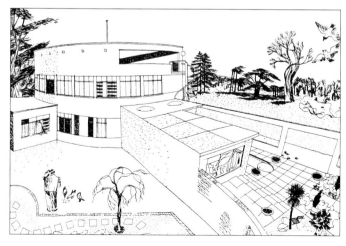

formal garden was as truly English as the much-touted landscape garden of the eighteenth century—and in the remainder of the book he tries to substantiate just that claim. Again surfaces the use of architectural order to distinguish the garden from the rude country-side, while simultaneously fashioning a bond between the two.

The century turned, England was profoundly affected by the Great War, a turn in mercantile practice, an increased reliance on technology, and the beginning of the dissolution of empire. New art forms such as cubism and surrealism changed the course of painting, the so-called International Style in architecture challenged the establishment, and atonality and serial composition influenced the course and sound of music. While historically the bastion of the most cherished, and thus most conservative beliefs, even garden design began to feel the pressures of the zeitgeist beyond its bounds. In the late 1930s, a Canadian-born, but English-trained, landscape architect named Christopher Tunnard earnestly attempted to formulate a statement about modernism in the garden. Tunnard had been given a boost by some still-born attempts to produce cubist-inspired gardens in the 1920s and early 1930s, what Dorothée Imbert has termed the *jardin tableau* or *jardin objet*, that is either the "garden as picture" or the "garden as object."[26] Plants had little to do with these gardens, except as sources of color, and on the whole they were rather static compositions despite highly wrought and zippy geometries. In fact, it takes no great leap of the imagination, or any great dip into the reservoirs of history, to see them merely as the updated shaping of parterres and other planted beds.

But these works were important in shaping the new view of the garden, perhaps more important for what they were *not*, than for what they were. They were not informal in the English manner, neither were they formal on the Italian or French models. And since they relied on the vocabularies of the new art, they must be modern. Or so the implicit thinking went.

Tunnard attempted to get beyond this limited notion of the garden as formal vocabulary, however. In a series of essays that ran in *The Architectural Review* in 1937 and 1938 (and collected later that year as *Gardens in the Modern Landscape*), he discussed, once again, the two principal traditions, the formal and the informal. But to these he added a third, new approach that he termed the empathic, which had asymmetry as its base [5-12]. "The neglected asymmetrical technique," he wrote, "embraces the delicate nuance as well as the bold impression.

It is symbolic of the plastic arts of our own time."[27] Like most of the writing in the book, the text develops its thesis only imprecisely. To clarify his argument, Tunnard advanced Japanese garden design as a model for the empathic approach. In Japan, he noted (somewhat naïvely), the garden is neither formal nor informal but an occult mixture of elements without perceived differences. There is no attempt to dominate nature, but "in the unity of the habitation with its environment," considering movement as an essential consideration of garden design. He ends this chapter by noting that: "The Japanese grasp of rhythm and accent in plant arrangements far excels our own, as does the marshalling of detail into significant and relevant patterns. It is the aesthetic conception which is the foundation of this virtuosity that must be allowed to seep into our artistic consciousness."[28]

That Tunnard sought to draw landscape design in tandem with architecture and other modern arts underscored his belief that landscape architecture must be a social art as well as a social science. While the implicit argument for an expression in accord with the times is buried in his text, he rooted the argument in reason, history, and problem solving. Indeed, in an article published during the war years, Tunnard proposed that "The right style for the twentieth century is no style at all, but a new conception of planning the human environment."[29]

It would be another twelve years, and after civilization had suffered yet another cataclysmic war, before the next major statement of landscape modernism appeared. Garrett Eckbo's 1950 *Landscape for Living* rejected the garden and the park as providers of mere visual pleasure, as the settings for horticultural display alone, or as the loci of the stylistic battle between the formal and the informal.[30] Instead, a landscape sited the interaction of people and place; landscape architecture—exterior spatial design—was the purposeful formation of that interaction. The book was part of an assembly of quotations by others, part recycling of earlier publications, and part an attempt at synthetic thought about the landscape [5-13]. If we are to produce landscapes for living today, Eckbo argued, we must look far more holistically than garden-makers had recourse to, or had needed to, in previous eras. Ecology was a major concern, but humans rightly occupy their position on Earth along with other living systems. Nature was not privileged by the author, who was no blind preservationist; instead, he asserted that man must design in accord with nature to create landscapes addressed to contemporary dwelling. It was high time, he said, to dismiss that fatuous opposition of formal and informal and instead look to the garden and landscape as a setting for living in harmony with the natural world.

Tunnard never reached a deep understanding of Japanese landscape design, turning his attention toward broader issues of city planning in the late 1940s. And he never encountered one particular aesthetic concept that could significantly have informed the making of modern landscape design. This was the practice of mixing within a single garden, elements of varying scales and varying formalities as relationships among consenting equals.[31] The simplest translation into English of the Japanese words *shin-gyo-so* might be "formal, semi-formal, and informal" orders, although typically, far more is implied in their implementation. Floral and calligraphic compositions—and even gardens—may be qualified as "formal" or "informal" [5-14]. But within the garden, it is not the existence of these categories as independent phenomena alone but their interrelations and juxtapositions—embedded one within the other—that have served as a generative force behind much of Japanese art and environmental design, especially during the Edo period (1603–1867).

It is believed that the terms shin-gyo-so were first used to characterize styles of calligraphy. Like the Japanese ideograms themselves, the shin style derived from Chinese orthography and was utilized in monumental inscriptions, formal documents, and name seals (hence, in China, it was termed Seal Script). Shin suggests formality: slow, somber, and considered fabrication; written forms chosen to conjure a feeling of dignity and substance.

As the brush moves more quickly, perhaps with greater acquired intuition than conscious thought, its trace assumes a different guise. The gyo style is a semi-cursive form in which the individual characters tend to merge together and assume a greater identity as a conglomerate than as individual markings [5-15]. Gyo involves the personalization of the broader cultural norm: an individual interpretation of a basic form. In the so, or "grass" style, the writing becomes so fluid it may not even be legible without some prior idea of the content: one can read the message only if one knows what the message might be. At this point the aesthetics of feeling transcend linguistic clarity, and written markings convey a meaning of their own.

In addition, the combinations of modes not only distinguish and define the type, but illustrate the power that derives from simultaneous contrast. These oppositions offer the mind a mental lens that perceptually zooms through the degrees of formality, its interest arrested only momentarily. The irregular stepping stones of the Katsura Villa, for example, play their roughness against the smooth, straight edge of an adjacent stone plank. Or a constellation of square stones overlays linear stripes of gravel. The contrast here is not only visual; to a garden visitor clad in kimono and wooden clogs, rough and/or irregularly placed stones necessitated careful footwork and continued consideration in order to negotiate movement through the space [5-16]. Each step required at least a glance at the succeeding step, while looking up after each considered pace unveiled a new view. Conversely, one could manipulate the placement of regularly shaped and surfaced stones as a means of effecting relatively easy movement while still maintaining a gyo aesthetic.

Within the formal areas, irregular elements such as natural stones within the shin composition of a path acquire heightened prominence.

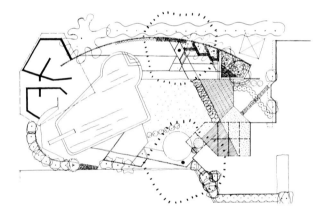

Through contrast, the formal parts invoke the softness of the non-geometric unit, for example, at Ryôan-ji, where the straighter edges of some stones contrast with rougher profiles of others, or the field of stones as an entity plays against the rectangular enclosing frame [5-17]. At the seventeenth-century gardens at Daichi-ji, in Shiga Prefecture, a shaped mass of green azaleas set against natural hillside planting dominates the temple's inner garden, itself framed by the formal architectonic structure. Periodic shearing maintains the hedge as a gently undulating wave, anchored at its ends by slightly more emphatically stated terminals. Nestled inside the soft curve of the hedge, the azaleas are cut into cubic masses that read as hardened forms within the soft vegetation [5-18]. The contrast between the geometric and curvilinear clipping is subtle and yet decisive,

[5-13]
GOLDSTONE GARDEN.
BEVERLY HILLS,
CALIFORNIA. 1948.
GARRETT ECKBO/
ECKBO, ROYSTON AND
WILLIAMS.
Plan.

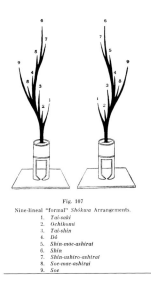

Fig. 107
Nine-lineal "formal" *Shōkwa* Arrangements.

1. *Tai-saki*
2. *Ochikomi*
3. *Tai-shin*
4. *Dō*
5. *Shin-mae-ashirai*
6. *Shin*
7. *Shin-ushiro-ashirai*
8. *Soe-mae-ashirai*
9. *Soe*

although the overriding identity of the hedge as a singular form pre-dominates.

Contrasting formalities create visual or conceptual richness in many fields of Japanese design. Kimono textiles overlay isometric grids with repetitive or abstract patterns that are in turn embroidered with floral motifs. The layering of surface design reflects an understanding of shin-gyo-so: the levels of ornamentation remain distinct, and yet reveal their intensity through differentiation. Each contributes to the awareness and comprehension of the other, yet each possesses its own discrete identity.

In actuality, the interweaving of shin, gyo, and so could be taken as an aesthetic philosophy, a means to analyze a completed work (although

[5-14]
FORMAL FLOWER ARRANGEMENTS.
[from Alfred Hoehn, *The Way of Japanese Flower Arrangement*, 1937]

[5-15]
WRITING IN THE *GYO* MANNER.
[From a teaching guide to Noh theater, twentieth century]

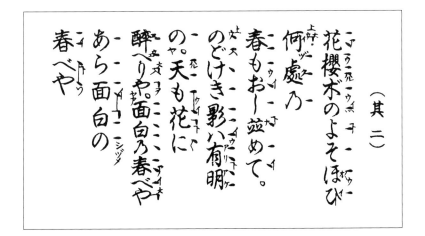

（其二）

花櫻ホのよそほひ
何處乃
春もおし並めて。
のどけき影八有明
の。天も花に
醉へりや面白乃春べや
あら面白の
春べや

this suggests a perhaps inappropriate application of Western rational-ism to Asian subjects), or more pragmatically as an overt vehicle for design. In all probability, plays of contrasts among the smooth and rough, the straight and irregular, the natural and polished, were endemic to the Japanese aesthetic sensibility. Their codification thus came after the fact, as a vehicle for disseminating that sensibility to a greater audience in place and time.

While the play of mixed orders was elevated to a refined art in Japanese garden design, it was found in other garden traditions as well, if not in quite the same way. From at least as far back as the sixteenth century, Western garden-makers have also addressed the question of landscape orders and the symbolic systems that those orders may represent. Composition remains a tool of intention, and like the forms that are ultimately perceived, both values and ideas are embedded within any structuring device. In Renaissance Italy the natural, or at least naturalistic, *bosco selvatico* complemented the formality of the villa's principal garden. And in French formal gardens such as Versailles, wooded areas or a romantic and later fragment such as Le Hameau, were fashioned as welcome respite from the rigid etiquette of court life embodied in the predominant axis and symmetrical disposition of the main garden [5-19]. Even the grotto of the landscape garden could be read as a fragment of naturalism set in contrast to the pol-ished architecture of the classical temple.

Although both Japan and Europe have shared the practice of combining systems of order, their respective manners for resolving those mixtures have varied considerably. Endemic to Western thinking, garden forms were typically conceived in terms of contrasting systems, balancing two distinct, if equal, poles on a binary gradient. The orders tended to be juxtaposed and complementary. The wooded bosco selvatico thus addressed the formal composition of pools and parterres; but the two zones remained somewhat segregated. At best, one could perceive the muted presence of one order within the other, for example, follies or statuary set among an otherwise naturalistic setting of clumped trees and plantings. Within the architectural framework, the vegetal deco-ration or motif might soften the masonry surfaces, as the trompe-l'œil trellises visually dissolved the confines of the garden room at the Villa Falconieri in Frascati or an arcade at the Villa Giulia in Rome.

In "Of Other Spaces: Utopias and Heterotopias," Michel Foucault established two principal classes of order.[32] The power of homotopic

order derives from the coherence between the parts and the whole. From the macro to the micro scales, a continuity pervades the complete entity; at any point the correlation between the part and whole is apparent. Homotopic order governs the composition of classical landscapes and architecture.

In formal gardens such as Vaux le Vicomte, and even more naturalistic parks such as Petworth, a homogeneity in thinking, elements, and composition directs the plantings and their positioning. The relationships of part to part, and part to whole, govern the design of the totality. The designer intends coherence, although allowing controlled variety in the constituent parts. While it is admittedly more difficult to acquire a static regularity using living materials, the order that structures their arrangement can be propelled by the urge toward homotopia. In these gardens coherence clearly dominates divergence.

Heterotopic order, on the other hand, results from collision. It celebrates disparity, and the fracture and collage of elements of varying formal properties. Order appears to derive almost by accident. While the analogy is perhaps strained to some degree, the Japanese utilization of shin-gyo-so tends toward the heterotopic. Although a certain coherence directs the limits of the design, whether the cut of the kimono or the architectonic frame that encloses the temple garden, the aesthetic power of many Japanese gardens ultimately derives from the simultaneous contrast of juxtaposed elements. By nesting one order within the next, the resulting garden compositions quietly seethe while simultaneously evoking a sense of stasis.

The power of garden design as visually perceived, it seems to me, rests in the balance of the garden's elements and the collisions and transitions among them. There is no formality in isolation—or informality in isolation—as there is no concept of nature free from a concept of culture. It would seem there is much to learn from Japanese garden art in its ability to resolve differences through juxtapositions: each order given its own identity, with each augmented by the presence of its seeming antithesis. Never is the question of formal and informal one of simple opposition, or a simple choice of one over the other.

Erase distinctions between orders, or play on the amalgamation of varying systems.

Hold, most of all, that the desired end justifies any appropriate means —whether that desired end is a rich and pleasurable experience, or a richly contemplative experience. That desired end also justifies any appropriate order, whether formal; informal; both; or none of the above.

Originally published in *Studies in the History of Gardens and the Designed Landscapes*, Volume 18, Number 2, April–June 1998.

Notes

1 This paper is a revised version of a lecture presented at the symposium "Formalism in the Garden," organized by Eric Haskell, and held jointly at Scripps College and the Huntington Library, 11–12 April 1997.

Sir Kenneth Clark once commented that turning lectures into essays was a form of literary suicide, given that verbal performance can mask a multitude of inaccuracies, weak points, and scholarly lacunae. The emphasis here, however, is placed on general ideas rather than universal applicability. To some extent I must request the indulgence of the reader who knows of specific examples that might severely strain or disprove my observations.

2 These age-old and rather tired distinctions between form and content would be dismissed by critics these days, but they seem to persist in the dictionary and in the collective conscious.

3 We rarely question, if this be true, why nature isn't good enough just the way it is; nor do we ever question the inherent presumption of our creating landscape settings that appear to be the work of nature or God. Nature, however, may be taken more abstractly, as evidenced in the biomorphic manner that pervaded garden design in the 1940s and 1950s. In some ways, biomorphism could be read as a confluence of human and natural systems, based on natural forms but reconfigured through human thought.

4 For some further commentary, see Marc Treib, "Traces upon the Land: The Formalist Landscape," *Architectural Association Quarterly*, Volume 11, Number 4, 1979, pp. 28–39 [included in this volume].

5 In fact, the opposition of formal and informal still pervades Western histories and theories of garden design (Humphry Repton's pro-nouncements on the selection of appropriate architectural style, for example). Its frequency once prompted the editor of *Studies in the History of Gardens and Designed Landscape*, John Dixon Hunt, to exclaim that the posed dichotomy was so contrived he wished the two terms would never cross his eyesight, and certainly his doorstep, ever again. Alas, here it is, one more time.

6 Jean-Jacques Rousseau, *Julie ou la Nouvelle Héloïse*, Judith H. McDowell, translator, University Park, PA: Pennsylvania State University Press, 1987, pp. 304–312 *passim.*

7 R. L. de Girardin, *De la composition des paysages* (An Essay on Landscape), 1783, Daniel Malthus, translator, reprint New York: Garland Publishing, 1982, pp. 3–4.

8 Ibid., pp. 5–6.

9 See Dora Wiebenson, *The Picturesque Garden in France*, Princeton, NJ: Princeton University Press, 1978.

10 In these contexts—and even in the most formal garden—topiary could be taken as an ambassador between the royal courts of architecture and vegetation, having geometry as its father and nature as its mother (or vice versa).

11 Criticism of Brown's designs cited the landscape gardener's planting of tree clusters in places where they would never grow in that form—for example, on the tops of hills when the lower-lying surrounds were open fields. Brown used the clumps to create closure and space and to dramatize the vista. In many respects, the criticism of Brown's manner is just, and it reveals the problems inherent in the use of elements or units true to nature, but employed in an order that clearly reveals the human mind.

"As Sir Uvedale Price said of Brown and his clumps of trees, 'While Mr. Brown was removing old pieces of formality, he was establishing new ones of more extensive and mischievous consequence.'" Reginald Blomfield and F. Inigo Thomas, *The Formal Garden in England*, London: Waterstone & Co., 1895, p. 14.

12 For an excellent investigation of Englishness in relation to garden design, see Anne Helmreich, "Contested Grounds: Garden Painting and the Invention of National Identity in England, 1880–1914," PhD dissertation, Northwestern University, Evanston, IL, 1994.

13 William Robinson, *The Wild Garden, or the Naturalization and Natural Grouping of Hardy Exotic Plants, with a Chapter on the Garden of British Wild Flowers*, 1870, reprint, London: Century Hutchinson/The National Trust, 1983.

14 Ibid., pp. 7–11, *passim*. There is more than a suggestion of an imperialist attitude at work here.

15 The illustrations by Alfred Parsons were part and parcel of Robinson's argument; like the text, they rarely present more than a cluster of species or a localized view.

16 Personally, I find this muddying of distinction between noun and verb more problematic for landscape studies than the traditional formal/informal enigma.

17 Gertrude Jekyll, *Wood and Garden*, 1899, reprinted in Penelope Hobhouse, ed., *Gertrude Jekyll on Gardening*, New York: Vintage Books, 1985, p. 23.

18 Her use of "drifts"—rendered as lozenge shapes on her detailed planting plans—softened the edges between species and more thoroughly their coloration than the simple juxtapositions of the historical planted bed.

19 Ibid., p. 213.

20 Ibid., p. 22.

21 Blomfield and Thomas, *The Formal Garden in England*. The book is available in a reprint by Thames and Hudson, 1985.

22 Ibid., pp. 2, 11.

23 Blomfield cites the remark of a "witty Frenchman": "Rien n'est plus facile que de dessiner un parc anglais; on n'a qu'à envirer son jardinier, et à suivre son trace." Ibid., p. 7. The English architect associates the meandering line with a lack of rational process and to some extent, propriety.

24 Ibid., p. 19.

25 Ibid., p. 20.

26 See Dorothée Imbert, *The Modernist Garden in France*, New Haven, CT: Yale University Press, 1993.

27 Christopher Tunnard, *Gardens in the Modern Landscape*, London: The Architectural Press, 1938, pp. 81–92 *passim*. Tunnard's connection to Japanese aesthetics seems to have been the noted potter Bernard Leach, who spent the 1920s and 1930s working both in Cornwall and in Asia.

28 Ibid., pp. 91–92.

29 Christopher Tunnard, "Modern Gardens for Modern Houses: Reflections on Current Trends in Landscape Design," 1942, reprinted in Marc Treib, ed., *Modern Landscape Architecture: A Critical Review*, Cambridge, Mass.: MIT Press, 1993, pp. 159–165.

30 Garrett Eckbo, *Landscape for Living*, New York: Duell, Sloan & Pearce, 1950, especially pp. 12–20.

31 Tunnard did not visit Japan until 1960, long after his immediate involvement with landscape architecture. I have developed the ideas in this section more fully in "Modes of Formality: The Distilled Complexity of Japanese Design," *Landscape Journal*, Spring, 1993, pp. 2–16.

32 Michel Foucault, "Of Other Spaces: Utopias and Heterotopias," 1967, translated and reprinted in Joan Ockman, ed., *Architecture Culture 1943–1968*, New York: Rizzoli International, 1993, pp. 419–426.

6

Must Landscapes Mean? Approaches to Significance in Recent Landscape Architecture

1995

I.

During the last decade, the amount of writing purporting to address meaning in landscape design has grown impressively.[1] Landscape architects now write of their attempts to imbue designs with significance by referring to such conditions as existing natural forms or to the historic aspects of the site. Cultural geographers, calling upon a collective body of study that now extends back well over half a century, interpret ordinary landscapes by first looking at the world around them; in their eyes, meaning congeals in setting, dwelling, and use— and not alone from the designer's intention.[2] Historians of gardens and landscape architecture tell us of those makers of places past who tried earnestly to create landscapes in which meaning would be evident and understood. At times relying on iconography and inscription, the creators of these gardens and parks sought to convey to the visitor a mental as well as a sensual impression. Within the garden confines, the visitor would take pause, and perhaps ponder the meaning of existence, or at least his or her part of it. Since the visitor, owner, and maker tended to share class and culture, intelligible communication was feasible.

These are only a few examples of the interests that have surfaced in the last decade and that have filled the pages of numerous publications. Principal among them, *The Meanings of Gardens*, edited by Mark Francis and Randolph Hester, Jr., in 1989, collected a series of essays that ranged in topic from religion to pop culture, from sex to pets, and geographically from Israel to Norway.[3] In the book, authors drawn from diverse disciplines questioned the significance of the landscapes we create; there were no generic conclusions although the essays were somewhat neatly arranged under the headings of idea, place, and action. In a 1988 essay titled "From Sacred Grove to Disney World," Robert Riley also tracked the search for meaning—and its removal over time—and concluded: "Gardens have been a locus of meaning in many cultures, but not in modern America."[4]

What are we to make of these renewed efforts to discern meaning in landscapes? Is it really possible to build into landscape architecture a semantic dimension that communicates the maker's intention to the inhabitant? If so, by what means? In addition, *should* we try to reveal meaning in environments, and if so, why? Where does the audience enter the process? Admittedly, this is notoriously treacherous territory and every author begins—and often ends—by hedging his or her bets. Laurie Olin stressed the "daunting" task of defining meaning and

[6-1]
VIET NAM MEMORIAL.
WASHINGTON, D.C.
1982+. MAYA LIN.
Much of the controversy over the memorial's design stemmed from the public's difficulty in finding meaning in an abstract, that is non-figurative, form. The power of the 55,000 names of the fallen engraved on the walls was inescapable, however, directing the reading of the monument as if captioned. To many, significance is commonly associated with identifiable representation, however, and a figurative statue of soldiers was later added to the grounds.

suggested that there were two broad categories in which the term was positioned. The first he termed "natural" or "evolutionary": "Generally these relate to aspects of the landscape as a setting for society and have been developed as a reflection or expression of hopes and fears for survival and perpetuation."[5] More simply stated, significance accrues through use and custom. Olin's second category, and the arena in which most designers operate, concerned synthetic or invented meanings, and it is these to which he devotes most of his essay and criticism.[6] My own effort will probably differ only slightly from that of almost every previous writer in that I will attempt to discuss the question of significance without precisely defining it.[7] To some degree this lacuna is problematic, in other ways it is may not be so troublesome.[8] I'd like to think that we can discuss the meaning of meaning in landscape without a definition applicable to all landscape circumstances, and I will operate under that premise. We can at least establish a broad theater in which meaning is taken simply as an integral aspect of human lives, beyond any basic attachment to the land through familiarity. Meaning thus comprises ethics, values, history, affect, all of them taken singly or as a group.

We could first try to establish *why* the pursuit of meaning has resurfaced at the close of the twentieth century. One reason might be the rejection of history, and all the baggage it carried, by those formulating a modern(ist) American landscape design in the late 1930s. Unlike their European colleagues, who continually confronted history in the world around them, American designers often started with a relatively clean slate. James Rose and Garrett Eckbo, among other writers, aggressively challenged the value of history as a lexicon of styles or typologies to be unquestioningly applied to contemporary problems and projects. Like their architectural contemporaries, they looked forward to solving problems of open space and form, and not backward to any book of given solutions. The received body of historical landscape architecture was taken as meaningless because its significance belonged to other places and other times.[9]

Rose, probably borrowing from the Canadian/Englishman Christopher Tunnard, argued for what he termed a "structural" use of plants: vegetation selected for a given climatic zone, but configured to create spaces to be used from within rather than to be viewed from without.[10] A continuing theme in Eckbo's writings well into the 1950s was the condemnation of the axis, which had "run out of gas in the 17th century."[11] Like Rose, Eckbo envisioned an enriched landscape configured for use, rather than one restricted to a linear spatial structure based on formal concerns alone.

There was little or no discussion of meaning in these writings, as there was—quite remarkably—no argument for any specific vocabulary. Significance derived from forms and spaces appropriate to their use and times; meaning was a by-product, or so the text implied. Although the zigzag was a popular feature in the gardens of Eckbo and Thomas Church, and the biomorphism of Jean Arp and Isamu Noguchi informed much postwar California garden design, no published texts connected these idioms with either modern art or the modern era—or argued for their significance.[12] In fact, very little was written specifically about syntax—that is the relationship between the elements—much less about semantic production.

Landscape writings of the period paralleled—almost always with a bit of a time lag—discourse on modern architecture. Sigfried Gideon, the central theorist for what came to be termed the International Style, rationalized the new architectural vocabulary by setting it against spatially vital architectures past.[13] The modernist art critic Clement Greenberg saw painting first and foremost as marks upon a canvas and found its culmination in non-objective works that stressed the flatness of the canvas; Gideon saw in modernist building the culmination of architecture as space.[14] In so doing, he actually recast history to accord with a twentieth-century vantage point. In anthropological terms, he was etic rather than emic, that is, looking at the subject from beyond its cultural limits rather than on its own terms. While a vast repertoire of Western architecture had accumulated over time, to Gideon its quest had ultimately been spatial rather than stylistic, and as such it reached a fruition in the modern era. Because he found space more central to architecture than either iconography or human affect, Gideon was more focused on architectonics (i.e., on architectural syntax) than on semantics. Or perhaps he regarded as synonymous significance and the means of spatial production. Garrett Eckbo's *Landscape for Living* of 1950 provided the modernist argument with its text and laid out the concerns and parameters for modern landscape architecture.[15] More fully developed in breadth and depth than earlier writings by either Tunnard or Rose, Eckbo reinforced the need for reflecting time and place and human presence in landscape architecture: but there was no discussion of what it meant.

In many ways, the next major ideological and highly polemical tract was Ian McHarg's *Design with Nature*, published in 1969. Focused on the evolving study of natural ecology, and rooted in landscape management, McHarg cited the natural world as the only viable source of landscape design. His text provided landscape architects with sufficient moral grounds for almost completely avoiding decisions of design—if design be taken as the conscious shaping of landscape rather than its stewardship. No talk of meaning here, only of natural processes and a moral imperative.[16] Laurie Olin, among others, has pointed out that design decisions normally derive from a greater complexity of factors than those of ecology alone, among them social and cultural issues including aesthetics and cautions: "[T]his chilling, close-minded stance of moral certitude is hostile to the vast body of work produced through history, castigating it as 'formal' and as representing the dominance of humans over nature."[17] McHarg mixed science with evangelism—a sort of Eco-Fundamentalism as it is sarcastically known by some parties—taking no prisoners and allowing no quarter.

The McHargian view was focused to the point of being exclusive, conflating two rather different arenas of landscape intervention/modulation as if they were one.[18] To manage a region without thorough "scientific" investigation and analysis would be fatuous, if not dangerous. Viable design begins with purposeful study and analysis. But the planning process rarely requires the active form-making that is central to landscape architecture. Reams of analysis and overlays will establish the parameters for making a garden for a suburban backyard, but they will hardly provide the design. McHarg's method insinuated that if the process was morally and analytically correct, the form would be good, almost as if an aesthetic automatically resulted from objective study.

Presumably, meaning would accompany the resulting landscape. The 1960s and the 1970s were dominated by attempts to rationalize the practices of architecture and landscape architecture, giving favor to social utility rather than the pursuit of form and/or significance. By the end of the decade, however, the limits to this way of thinking, coupled with an emerging desire by younger landscape architects to again become visible, began to generate a reaction to the anti-aesthetic and anti-semantic climates of the preceding decade.

Admittedly, this is a cursory explanation of a professional condition that derived from a complex series of interrelated factors. Landscape architecture is, after all, a part of a cultural, technical, and social milieu and as such is informed by a multitude of factors and considerations. But …

II.

During the 1980s, declarations of meanings began to accompany the
published photos and drawings of landscape designs [6-1]. At confer-
ences, landscape architects described their intentions, their sources,
and what they believed their designs signified. Some designers merely
claimed they were once again touching base with the vernacular matrix
in which High Style design was embedded. Martha Schwartz, for
example, re-examined the materials of the ordinary landscape and the
typologies of the small private garden and the shopping center. George
Hargreaves spoke of a perceptually complex space at the 1984 Harlequin
Plaza in Inglewood, Colorado, although he shied away from making
direct claims about its meaning(s) [6-2]. The emerging generation of
designers displayed a new interest in making form; and many of them
claimed that these new forms would be meaningful. Landscape architecture from these two decades might be assigned to one or more of
five roughly framed approaches, and by extension, to a striving for
significance: the Neo-Archaic, the Genius of the Place, the Zeitgeist,
the Vernacular Landscape, and the Didactic.

A sort of primitivism constituted one attempt to retrieve that which
had been lost at some unspecified point along the way to modernity.
Borrowing from schools of art that ranged from the body works of
Ana Mendieta to the stone markings of Richard Long to the theories
of entropy proffered by Robert Smithson, landscape architects began
to reconfigure the land in a manner we could term *Neo-Archaic*. Whether
the landscape architects referred directly to neolithic sources, or only
to the sculptors who had drawn upon them, is impossible to determine.
Perhaps they tapped both resources. But in neighborhood playgrounds
and in suburban office parks, one began to encounter hills coiled with
spiral paths, cuts in the earth aligned with the rising or setting sun
(or the summer solstice), circles of broken stone, and clusters of
sacred groves. Granite stelles evoking the stone circles of ancient
Scandinavia—or was that England's Salisbury Plain or Easter Island?
—appeared in backyards and plazas. Myriad versions of Jai Singh's
eighteenth-century astronomical observatories at Delhi and Jaipur
popped up like mushrooms, including one reinterpretation in a fine
garden by the master Isamu Noguchi [6-3].[19] One can almost hear
designers saying, *sotto voce*: "If they meant something in the past (of
course, we have to like them as forms...), then they will mean something again to us today." Gary Dwyer's proposal to link the two sides
of the San Andreas Fault in California with criss-crossed topographic

[6-2]
HARLEQUIN PLAZA.
ENGLEWOOD,
COLORADO. 1982. SWA /
GEORGE HARGREAVES.
The post-modern
challenge to the
ecological and
romantic
approaches.

band-aids curiously developed from the Ogham writing of the Celts is extreme to be sure—and rather difficult to support with rational argument—but it was not at all that bizarre in the context of contemporary projects.[20] As Catherine Howett once aptly phrased it: "By the early 1980s, every landscape architect student project had been equinoxed to death."[21]

If archaicism was one school of semantic creation, the worship of the *Genius of the Place* marked a second. Alexander Pope had enjoined Lord Burlington to consult the spirit of the place as a means of rooting landscape design in a particular locale. A garden was not a universal concept to be applied without inflection upon all sites. Instead, the garden revealed the particularities of its place as well as the profundity of the garden's idea. Long driven underground by the onslaught of urbanity, suburbanity, and modern technology, the genius was a bit hesitant to re-emerge into the twentieth-century sunlight and, as a result, came out squinting. A renewed cult figure, the genius—or what was left of him or her—could be consulted in many places in only a desultory way, since "the place" had been so disturbed by centuries of industrial and residential development. While writers such as Christian Norberg Schulz based their discussion of the genius and place in the phenomenology of Maurice Merleau-Ponty, and others decried the rise of an endemic placelessness, designers often adopted superficial approaches to connect human inhabitants and their landscape setting.[22]

History became an image to be dusted off and applied to any current proposal as a means of validation. In a glance over the shoulder of history, the tiny urban park was planted with prairie grass to show what vegetation had once thrived there. Like the caged animal in the zoo, however, an urban prairie is hardly a prairie at all—it is an urban garden planted with unmown grass and little else. At best, it has been reduced to a sign for what had been. Since the frame for reading—that is to say its context—had been so drastically altered, the subject of view is not easily understood by contemporary citizens as a reference to the past. The grass that so magnificently sheathed the prairie has been reduced from an inherent and meaningful component of early settlement to a design, or at best, a museological feature. Meaning arrives on the face of a plastic or metal plaque that credits the designer, the sponsoring body (usually a benevolent foundation for Green America), and of course, the mayor in office at the time. Still, most passersby wonder quietly to themselves: "When are they going to cut that lawn? I'm sure there are rats and Lord knows what else living in it. And they should water it; it looks dead."[23]

The presence of the genius is a bit more obvious in undisturbed land, but there is precious little of that around these days, as s/he has hardly been left unaffected by changes in atmosphere and climate. Still, the genius provides major support for landscape design and its rationalization today. Technically, studies of vegetation, hydrology, soil conditions, and the like are indeed the basis of design; but do these suggest a significant form for the design? If there is a stand of oaks, do you plant more oaks? Or should the stand be complemented by another species that even to the untrained eye appears to be foreign to the site?[24] So much of landscape architecture in the past has been

[6-3]
NEO-ARCHAICISM:
CALIFORNIA SCENARIO.
COSTA MESA,
CALIFORNIA. 1984.
ISAMU NOGUCHI.
Beneath the seeming
sculptural abstraction
of this plaza/garden is
a program to evoke the
various ecological zones
of the Golden State. The
wedge form of the water
source, suggestive of
the Sierra Mountains,
recalls the Indian astro-
nomical observatories
at Jaipur and Delhi.
Noguchi has attempted
to tap the genius of the
place as well as archaic
associations. One could
also assign this work,
perhaps with equal
merit, to the Didactic
manner.

created to *overcome* what the genius of the place offered the "unimproved" land—for example, by bringing water to the desert or by constructing conditioned enclosures to grow oranges in colder climates—that it is obvious that his/her ambiguous advice can be interpreted rather freely. In instances such as the Patio of the Oranges in Seville the human contrivance of irrigation was elevated to an art form, creating a garden of exceptional pleasure, refinement, and calm. Needless to say, this beautiful and beautifully verdant forecourt was not conceived as a xeriscape that relied upon native plants; admittedly, it was collective and religious, rather than an anonymous private, vernacular garden. But this courtyard—like other pieces of greenery and water in arid climates—nevertheless illustrates that while one should consult Genius & Company, one needn't accept the advice in precisely the manner it was given. Like any consultation, the information must be evaluated and some decisions need to be made, including those of form [6-4, 6-5].

Buried within this approach to shaping the landscape is the belief that reflecting a pre-existing condition creates a design more meaningful to the inhabitants. Why so? Many of these inhabitants weren't present on the planet at the time the land was pristine. I recently attended a project presentation that informed those gathered that as the principal concept for a natural preserve the designers and clients had recently restored the historical ecology in its original pattern. That they had also created a pond where none had existed—presumably as much for the visitors as for the birds that were to be lured to this reserve —was passed over without question. It is difficult to fault the good intentions of restoring disturbed wetlands. But why "restore" the original pattern when, in fact, the reserve today serves equally for human recreation and open space preservation? Is it because the "natural" pattern, masquerading as nature, is less open to question by client or visitor alike? Or could it be that the designers, somewhat defensively, do not believe they can improve upon the natural pattern to bring the landscape into greater accord with the new uses and the drift of the times? Or is it a conscious or unconscious harking back to received picturesque values? Does the genius really grant significance, or does he or she just point out the easiest path to follow, what in the zoological world is called a "target of opportunity"? [This will be discussed further in Part III below.]

Approach number three borrows from related disciplines, which suggests a belief in the zeitgeist (i.e., "the spirit of the times") as a determining

[6-4]
THE GENIUS OF THE PLACE: PATIO OF THE ORANGES. SEVILLE, SPAIN. SIXTEENTH CENTURY+.
The Genius of the Place is always open to interpretation—to accept the advice or to elevate it to another plane?

[6-5]
THE GENIUS OF THE PLACE: SUZANNE DELLAL DANCE CENTER PLAZA. TEL AVIV, ISRAEL. 1990. SHLOMO ARONSON.
Sharing a similar climate and addressing similar environmental constraints, this recent plaza design recalls the patios of the oranges at both Cordoba and Seville: pragmatic irrigation furnished the point of departure.

[6-6]
ZEITGEIST:
TANNER FOUNTAIN,
HARVARD UNIVERSITY,
CAMBRIDGE,
MASSACHUSETTS.
1984. PETER WALKER.
The rocks recall the
materials of Carl Andre's
Stone Field Sculpture;
their configuration, the
work of Richard Long.

[6-7]
*STONE FIELD
SCULPTURE*,
HARTFORD,
CONNECTICUT. 1977.
CARL ANDRE.
Taking a clue from the
Genius of the Place,
Andre selected boulders
that represented the
types of stone found in
the areas surrounding
the city; they are
arranged in a rigid,
non-naturalistic, grid.

force for any aspect of contemporary culture. If artists, and the battery of cultural critics that support and explain their work, have produced a body of work deemed illustrative of the spirit of our times, then land-scapes designed with contemporary art-like elements must share that significance. Such an approach intersects at times with the Neo-Archaic, particularly in recent years when a new regard for pre-history has informed at least one major strain of art-making.[25]

The boulders that comprise Peter Walker's Tanner Fountain at Harvard from 1984 bear a striking resemblance to those Carl Andre had neatly arranged in his *Stone Field Sculpture* in Hartford some seven years earlier [6-6, 6-7]. Andre, in spite of his ultra-minimalist proclivities, had actually consulted the Genius in creating the work, choosing a range of stone types native to the surrounding area as the basic material of the installation. (Because the rocks had been removed from their native context, however, the populace required a written or verbal explanation to inform them of the fact.) Walker's stones, in contrast, are all more or less the same size and type, and their circular config-uration—like certain elements of his later IBM Solana, Texas, campus —also cite rather directly the work of sculptor Richard Long. Certainly an aesthetic transformation has been employed; neither the fountain nor courtyard design in any way constitute a plagiarized form. But much of their novelty and appeal—at least at the date of their initiation —derives from their seeming correlation with art forms of the times. From sculpture, the designer receives both the instigation of ideas and, to some degree, of validation. Landscape architecture becomes in the process a part of the ethos of the era, and its own identity as an art is confirmed. This is not a bad practice as it acknowledges the importance of regarding landscape architecture as a cultural as well as environmental practice.

Perhaps the most prominent recent example of the zeitgeist approach is the 1988 Parc de la Villette in Paris, won in competition by Bernard Tschumi. Bounded on one edge by the Périphérique [ring road], described by architectural historian Norma Evenson as the concrete moat that surrounds Paris,[26] the site was offered little by the Genius Loci, and a Didactic [see below] approach would have demanded a strong evocation of the site's history or even the reinstitution of the animal slaughter that once transpired on the site.[27] Instead, Tschumi used notions of cinematic sequence, and post-structural theories con-cerning the fragmentation of post-modern culture, as his sources for the park's design. The "outmoded" concept of park was supposedly dissolved by these new ideas, instead producing a design that effaced the boundary between city and park, and eliminated the hard line between built and green zones.

The drawings used to explain the competition design were brilliantly conceived, and included an exploded axonometric view that masterfully conveyed the design concept of point, line, and surface—a visual equivalent of a "sound-bite." Unfortunately, parks are rarely seen from the air, and even less frequently as exploded axonometrics. In fact, as a totality, the non-composition recalls too closely the bland and amorphous open spaces of Paris's *grands ensembles* (housing projects) of the 1950s and 1960s [6-8]. La Villette's red follies, while intriguing as investigations of architectural form, do little to energize the park's sensual appeal beyond the visual. Ultimately there is precious little of genuine, that is to say *experiential*, interest as landscape architecture on the site. Basically, the landscape comprises some lawn and some rows of trees.[28] The ideas used to conceive the park are rich and evocative; the experience on site is limited and somewhat spatially uninteresting, however. At what point does concept end and experience begin? Is an intriguing concept sufficient to create meaning in the minds of the beholders? What of the beholder not privy to the designer's convoluted explanation? The Parc de la Villette illustrates the problems that plague borrowing parallel ideas or forms from other disciplines, and the distortion that often accompanies translation. In this particular Parisian example, what has been written about the project is far more intriguing than what one encounters on site.[29] The ultimate success or failure of such landscape designs does not ultimately derive from their intellectual origins, but whether or not they "work" on their own merits as places and landscapes, without recourse to jargon and verbal explanations. One might also ask in the end: What pleasure do they provide?

Like architects such as Robert Venturi and Frank Gehry, landscape architects such as Martha Schwartz also look at the *Vernacular Landscape*. This is a hip glance at the Genius of the Place, of course, but the genius is culturalized and the selections suave. The vernacular is a rich source of materials and forms; after all, it constitutes the "real" world in which we dwell. But just as the meanings of historical landscapes are affected by re-framing, the vernacular landscape is inevitably transformed when borrowed by design professionals. And when vernacular elements reappear in High Style projects, they have semantically virtually nothing in common with their sources. They have been re-framed and recon-

[6-8]
ZEITGEIST:
PARC DE LA
VILLETTE.
PARIS. 1988.
BERNARD TSCHUMI.
Despite the trendiness
and photogeneiety
of the follies and the
intrigue of their archi-
tectural form, there is
precious little of
experiential interest
within the landscape
elements themselves.

stituted. The vernacular environment is treated by designers as a quarry for raw materials to be reconfigured and thus transfigured. The unselfconsciousness, the appropriate sense of the makeshift, and the accepted transience of vernacular building normally disappear along the way.[30] A glass gazing ball optically enlarges the confines of a small backyard private garden, while serving as a sign of neighborly propriety. When it is extracted from the backyard, repeated at length, and arranged in a grid, however, only the basic reflective properties remain unaffected [6-9]. Similarly, a concrete frog accompanying a cement deer and perhaps a gnome, are tender companions in an intimate garden setting. Multiplied by the hundreds and painted gold, they are no longer the common vernacular element they once were, but fodder for High Style designers. This is not to say they possess no merit of their own, they most certainly do; but the meaning is no longer vernacular. Like fine wine, significance doesn't travel very well, and wine *does* differ from grape juice.

The fourth approach to "constructed meaning" goes down the *Didactic* path. This is the one I have found most appealing, and one which has formed the only more-or-less stable leg of our own projects. In fact, it was the observation by a friend while examining a current project that made me realize that much of what we do is a somewhat desperate search for meaning in landscape.[31] The Didactic approach dictates that forms should tell us—in fact instruct us—about the natural workings or history of the place. This is related—as all the approaches are to some degree—to the Genius Loci school, but the Didactic is usually more overt in its intentions. Not only should we consult the Genius

[6-9]
THE VERNACULAR
LANDSCAPE:
INNOVATIVE
TECHNOLOGY.
HERDON, VIRGINIA.
1988. MARTHA
SCHWARTZ, LAND-
SCAPE ARCHITECT;
ARQUITECTONIA,
ARCHITECTS.
While the glass
gazing balls glitter and
brilliantly reflect the
building's curtain walls,
their meaning has
been fundamentally
transformed through
displacement and
multiplication. The
vernacular garden
element has been
reframed and recast
within the precepts of
the High Style.

about its basis, but our resultant project should render an exegesis on what the Genius told us.[32]

Curiously, we often try to restore what has been previously destroyed. Perhaps a stream long culverted and buried is restored to its "original" state (of course, it really isn't—everything has changed around it). One of the rules formulated by Joel Garreau in *Edge City* is that one names a place for the features that have been destroyed to make room for the new development.[33] Shady Hills Estates commemorates the trees that were cut to build the houses, and the natural undulations that were flattened to make construction less challenging; and incidentally, the houses are hardly estates. But like the photo caption, the name of the development directs our reading of the place and asks us to complete that which is missing in the picture. A design didactically conceived, like the photo caption, is both informative—possibly normative—and certainly directive. The "factual" base is intended to validate the designer's work.

A Didactic landscape is supposedly an aesthetic textbook on natural, or in some cases urban, processes. Alexandre Chemetoff's sunken bamboo garden at La Villette purposefully allowed the elements of urban infrastructure to remain, reminding the visitor that this small green respite was actually but a fragment of an urban agglomeration that to exist required massive amounts of servicing [6-10]. Water mains, sewer pipes, and electrical ducts crisscross the site; the retaining walls are constructed of pre-cast concrete elements commonly used to support the walls of adjacent sites during excavation for new construction. The landscape architect did not leave these elements of infrastructure untouched, however; the scheme itself developed as a give and take between the didactic exposure of services and its aesthetic complement in wispy green and gold foliage. Sculptors—who almost by definition are allowed to consider the aesthetic parameter in isolation—have also created places structured on the Didactic dimension. At the National Oceanographic and Atmospheric Administration in Seattle, for example, Douglas Hollis's *A Sound Garden* (1983) captured the wind to activate an environmental organ; the vanes aligning the field of erect pipes into the gusts added a visual signal of wind direction [6-11]. Here the presence of the wind received both aural and visual expression.

In these two instances, the work of landscape art addressed either natural or urban process with an assumption—which I have since come to suspect—that designs revealing these processes are both more viable and more meaningful. I don't think the answer is quite that simple. Didactic thinking provides a good point of departure for the work, but the success of the place ultimately hinges on the skill and care with which the design is made and on what it offers the visitor. Didacticism per se is not enough. (In these two instances the final success of the resulting works—and I do regard them as successful —did not depend on their Didactic aspects alone, but a host of factors that does not exclude siting, materials, aesthetics, and sophistication.)

And then there is the *Theme Garden*. It is curious to me how many people deride the world's Disneylands and other theme parks, and

[6-10]
DIDACTIC:
SEQUENCE IV, OR
BAMBOO GARDEN,
PARC DE LA
VILLETTE.
PARIS, FRANCE. 1988.
ALEXANDRE
CHEMETOFF.
Within this incised
ravine the necessary
elements of urban
infrastructure—sewers,
water mains, and
electrical services—
remind the visitor that
even seemingly natural
environments within
the city require
human support.

[6-11]
DIDACTIC:
A SOUND GARDEN.
NATIONAL
OCEANOGRAPHIC
AND ATMOSPHERIC
ADMINISTRATION,
SEATTLE,
WASHINGTON. 1983.
DOUGLAS HOLLIS.
The metal vanes that
keep the pipes aligned
with the air flow visually
reinforce the aural
presence of wind.

then propose Theme Gardens. A theme, in this context, constitutes a perceptually apparent idea used to fashion the garden's form. Roses, Mother Goose, the color yellow, or even electric light could all be used as themes and I imagine that all of them have already been used as such somewhere at some time. One could argue that the gamut of themes deriving from horticultural or environmental ideas or cultural borrowings are inherently more genuine than the contrived imagery of a theme park created in plaster or plastic, but they are themes nonetheless.[34]

A theme, it must be admitted, is not necessarily an argument for significance, but there is an underlying assertion of validity that accompanies any obvious concept. Even today, the landscape professional can accept a Chinese garden, for example that by Fletcher Steele at Naumkeag [6-12], or the copper tents at Frederik Magnus Piper's eighteenth-century Haga Park in Stockholm [6-13]. Perhaps we use the word "charming" rather than "beautiful" to qualify them. If well done, in fact, the effect of the pavilion or cultural borrowing is far greater than its semantic theme. It can be pleasant, calming, restful, stimulating in its own right; that is, it can affect us. Which tells us something about the experiential dimensions of the garden.

The white garden at Sissinghurst is a well-known example of color used as a subject, but the themed approach is widespread in time and place. The recently opened Parc André Citroën in Paris includes "black" and "white" gardens, although in both gardens green seems to be the predominant color that meets the eye. One could argue that restriction to a single color suggests a poverty of horticultural invention or an overly zealous pursuit of minimalism [6-14]. It can also, of course, create a garden of stunning beauty, employing incredible horticultural acrobatics and subtle chromatic mixtures even with a single color range.

Gilles Clément, the landscape architect responsible for a considerable section of the park, has also applied his idea of a "garden in movement" to one of its riverside zones. Here a score of wild flowers and grasses has been planted with little regard as to where or to which will survive. Paths through these meadows will be determined by human movement rather than by formal design; the paths will fix the traces of occupation and use. This Darwinian approach to park design—which joins the Didactic with the Theme with instructive aesthetic consequences— addresses both the social issues brought to the fore in the 1960s and aspects of urban ecology. While these parts of the park will evolve in terms of horticultural species—and over time run the risk of looking like a vacant lot—they suggest the human presence only through a relatively few wooden seating platforms raised slightly above the ground. The idea of replicating evolution to establish an appropriate urban landscape is engaging, although the form may not be attractive at all times. But do they mean anything to anyone today?[35]

III.

Is it really possible to imbue a place with meaning from the outset? It would seem that history tells us yes, if the users possess sufficient experience in common. For one, significance is culturally circumscribed and ultimately, personally determined.[36] If we examine a Chinese poem executed in ink on silk, as non-readers of the Chinese language we are denied access to the poem's linguistic dimension. Should we be un-initiated into Chinese calligraphy, and the propriety and taste conveyed by the chosen style, the marks will have even less meaning to us. Should we so choose, we can always appreciate the work solely for its formal dimension, of course, as fluid black marks on a white ground. It is obvious, however, that possessing linguistic abilities in Chinese would enrich both our understanding and our pleasure: the two-dimensional writing on the page would acquire multiple semantic dimensions.

The same is true of gardens. The uninitiated may or may not appreciate a dry "Zen" garden for its formal properties alone, for the pattern of its

[6-12]
NAUMKEAG.
STOCKBRIDGE,
MASSACHUSETTS.
CHINESE GARDEN.
1937, 1955.
FLETCHER STEELE.

[6-13]
HAGA PARK.
STOCKHOLM, SWEDEN.
COPPER TENTS. LATE
EIGHTEENTH CENTURY.
FREDERIK MAGNUS
PIPER.

[6-14]
THEME GARDEN:
PARC ANDRÉ CITROËN.
PARIS. 1992. GILLES
CLÉMENT, LANDSCAPE
ARCHITECT (FOR THIS
PART OF THE PARK).
There are a number of
theme gardens within
the park, among them
the garden of grasses
and the "garden in
movement." The theme
is probably more
important to the
designer's instigation of
ideas than the visitor's
perception of the park.

raked sand and the composition of its rocks; but the cultural meanings of the garden will be communicated imperfectly at best. The absence of many of the elements that say "garden" to members of foreign cultures denies access to meaning as the religious proscriptions deny physical access into the space.[37]

The Japanese dry garden offers a valuable case study for considering the construction of meaning. Japan's centuries of cultural homogeneity fostered an attitude toward simplicity as the compression of complexity (rather than its reduction or elimination, as it has been in the West). One could say, with perhaps only a little exaggeration, that until relatively recently a Japanese of a certain class/educational level could understand the intentions behind the making of the garden. He or she could appreciate the framing of the space; the non-geometric order within the enclosure; the quality of the rocks and their arrangement; the shaping of shrubs; the almost complete absence of brilliantly flowering species. Unless familiar with Zen doctrine, however, the site's significance as an embodiment of religious belief, and as possibly intended by the garden-maker, would remain beyond comprehension. And since Zen reflects continually back on the self for understanding and ultimately enlightenment, there is an implicit denial of meaning within the landscape itself. Instead, the garden as well as its care may stimulate individual contemplation; it can be seen as a vehicle for understanding the self rather than the place. The meaning of the garden is non-meaning. In Zen belief, the place bears no meaning per se, but it can perhaps evoke a call for meaning within the individual.

Allusions to worlds beyond the garden in place and time have appeared with some regularity in the polite traditions of landscape design in both East and West.[38] Replicas or recollections of Roman temples often appeared in the English landscape garden, for example. At Katsura Rikyu in Kyoto a small spit of water-worn stones was intended to cast the visitor's musings toward the peninsula of Ama-no-hashidate, long regarded as one of Japan's most outstanding shoreline landscapes [6-15]. The shorn bamboo-covered slope at Koraku-en in today's Tokyo, on the other hand, specifically invoked the Mountain of the Chinese Immortals. Unlike the abstract Zen landscapes, that were intended to summon a multitude of (ultimately personal) interpretations and associations, the aristocratic villa gardens often established intimations of legend and land. Meaning accrued from allusions to real or mythic geography outside the immediate landscape.

John Dixon Hunt has cogently argued that the world of the English landscape garden, like many garden traditions before it, was a coherent system of signs devised to be legible to both maker and visitor.[39] Here the signs were made tangible: a temple based on a familiar Roman predecessor; a vale with mythological reference; an architectonic emblem of Englishness. References could be manifest in a landscape feature, a structure, or even a written inscription to reduce ambiguity. Although falling under the common heading of signification, they actually concern two structures of meaning, differentiated in time. The first regarded the production of meaning used at the moment of the garden's creation and its effect(iveness) on the visitor. The second concerned the greater orbit of meaning that is part of the garden as an institution and semiotic constellation. "Gardens, too, mean rather than are," claims this garden historian. "Their various signs are constituted of all the elements that compose them—elements of technical human intervention like terraces or the shape of flowerbeds, elements of nature like water and trees—but they are nonetheless signs, to be read by outsiders in time and space for what they tell of a certain society."[40] Hunt also states—at first seemingly in contradiction with what he has written earlier in the essay—that even the most specific of references (probably textual ones) become time worn and lose their significance: "Castle Howard and Rousham provide excellent examples of garden experience we have totally lost. We no longer see a representation of English landscape; we just see it."[41]

Any symbolic system demands education for comprending both the medium and the message. One might understand, for example, that Diana was the goddess of the chase, and even know of her association with the moon, but still might have absolutely no idea why her likeness stands in the garden. Were we unaware of Louis XIV's self-association with the sun, would we not believe Versailles to be a glorious homage to cloudy France's sunshine lost or to Apollo himself? We may have lost the ability to read some or all of the original intentions, but we can still decipher the original garden elements on our own contemporary terms. That these two worlds of meaning mutate over time suggests that meaning is indeed dynamic and ever-changing.[42] It also suggests that the meaning with which the designer believes he or she is investing the garden may have only minimal impact in the beginning, and even less in years to come.[43] On the other hand, he or she does have power over the artifact and its immediate effect on the senses — and its potential to mean.

Communication theory tells us that the two parties in conversation must share a common semantic channel or there will be no real interchange; no communication. Can the garden operate as such a channel, and does the designer possess the power to create a significant landscape, especially given the multitude of communication channels in today's pluralistic world? When a society is relatively homogenized, the task is far easier because the designer shares the values and belief system of the people. Folk cultures produce places that are almost immediately communicative, and communicative over long periods. Because their connections between form and intention are understood within the culture and evolve only slowly over time, it is possible for the makers, the people, and the meaning of place all to remain in contact.[44]

The Woodland Cemetery outside Stockholm, designed between 1915 and 1940 by Gunnar Asplund and Sigurd Lewerentz, tapped into the religious and value systems of the Swedish Lutheran congregants. This landscape of remembrance has remained both meaningful to its parishioners, and appreciated by them, from the time of its realization. The triumph of the cemetery lies not only in its magnificent joining of architecture and landscape, and the modulated juncture of re-formed land with the existing pine forest, but also in its ability to conjure a sense of sanctity without relying on overt Christian iconography. Perhaps the power of this funereal landscape ultimately derives from an almost animistic feeling of pre-Christianity that addresses the forest, the land, and the heavens as a primeval setting [6-16]. Perhaps the design also tapped into something basic to Swedish religion and culture. It

[6-15]
THE AMA-NO-HASHIDATE PENINSULA, KATSURA RIKYU. KYOTO. SEVENTEENTH CENTURY.
Although using natural, unworked stones, the peninsula as a totality functions as an abstraction—a sign—for the true landscape found on the northern Japanese coast. Meaning is transferred from an actual scenic place in Japan to the constructed garden.

might still be possible to create a landscape equally attuned to its time and place today, when Swedish society is far more diverse. But it would be far more difficult to devise the forms and symbols that would resonate within the contemporary Swedish population in quite the same way.[45] Not that it was ever easy; but it was certainly easier earlier in the century. The communication channels are no longer so few, nor are the elements of the Swedish landscape so simple.

To summarize:

Can a (landscape) designer help make a significant place? Yes.

Can a (landscape) designer design significance into the place at the time of its realization? No, or let's say, no longer in most places.

When a society was homogenous and shared a common system of belief, when the symbolic system was endemic, when the makers of places operated unselfconsciously fully within the culture, it was possible.[46] But even then, meaning was enriched through habit and the passage of time. Given the fragmentation of contemporary American society, and especially with its current emphasis on difference, the concord necessary for instant meaning is—to say the least—deficient.

Since a commissioning body might include meaningfulness as a part of its brief, why commission a (landscape) designer?

[6-16]
WOODLAND CEMETERY. ENSKEDE, SWEDEN. 1915–1940. ERIK GUNNAR ASPLUND AND SIGURD LEWERENTZ. The architecture and landscape designs constitute an embodiment of shared cultural and religious values, and were relatively successful in creating a meaningful landscape.

Of course, there are the pragmatic aspects of design that can best be addressed by those with an education, technical knowledge, and experience. One also hopes that the landscape architect possesses equal skill in understanding people and culture as well as horticulture and form. Creating significant landscapes remains a quest of the profession, as well it should. But calling attention to Celtic inscriptions, solar alignments, the spirit of the place, the zeitgeist, the vernacular landscape, or even a didactic lesson in the derivation of form, does not create meaning. Providing signifiers is not the same as creating significance, although it may be one point along the path. To my mind significance ultimately lies with the beholder and not alone in the place. Meaning accrues over time; like respect, it is earned, it is not granted. While the designer yearns to establish a landscape that will acquire significance, pat symbols alone do not transform syntax into semantics, that is, tectonics into meaning.[47]

Familiarity and affect are not quite the same as significance, although they can serve as vehicles for its creation. To recall the site of one's first camping trip, or the park where the football championship was won, or even the flowers of one's family home ground establish associations among place, act, and form that cohere in landscape meaning. If these places were designed by landscape architects, all well and good. Meaning condenses at the intersection of people and place, and not alone in the form the designer's idea takes.

The design itself constitutes a filter that creates the difference between what the designer intends and what the visitor experiences. This is the difference between the intended perception and the perceived intention. Differences in culture, in education, in life experience, in our experience of nature, will all modify our perception of the work of landscape architecture. While this transaction between people and place is never completely symmetrical, we can circumscribe the range of possible reactions to a designed place. We cannot make that place mean, but we can hopefully instigate reactions to the place that will fall within the desired confines of happiness, gloom, joy, contemplation, or delight. This range of possible reactions, while tempered by cultural norms and personal experience, is still physiologically dependent on the human body. The limits of thermal comforts, the olfactory faculty, the capability to perceive chroma and natural process, and our basic size are characteristics shared by virtually every human inhabitant of the planet. Could we not start with these physical senses rather than with the encultured mind? Could we not make the place pleasurable?

IV.

In historical garden literature a considerable amount of text describes the pleasure of the garden, that is, its comfort, its delight, its sense of well-being. The pleasures of the imperial palace garden in Kyoto backdrop the rather limited action and plot development in the twelfth-century novel *The Tale of Genji*. Pleasure and its appreciation were so much a part of gardens in the past that we can well wonder why land-scape architects today seek significance rather than pleasure. Could it be that pleasure is trite, hedonistic, and ephemeral while meaning is deep and long lasting? Or perhaps pleasure seems to be too solitary an enterprise while meaning is taken as collective embodiment of values? Or is it that meaning is the dimension that distinguishes landscape architecture from mere "gardening"?

Roland Barthes argued that to read is to seek the pleasure of the text. He tells us that to provide pleasure: "The text must prove to me *that it desires me*" [italics in the original].[48] Knowledge, a magnificent use of language, plot, linguistic constructions—all contribute to the ultimate goal: the pleasure of reading. Can we not suggest that pleasure is one of the necessary entry points to significance (certainly horror would be another, as the sublime school once believed, but our quotidian world seems to provide enough of that)?

It seems curious to me that in most professional design publications, the aspect of pleasure is almost completely missing from the discourse while it thrives in popular gardening magazines and in seed catalogs. This is not to say that pursuit of pleasure is not a part of professional work; one assumes that park design, for example, is to a large degree predicated upon the contented use of its grounds. But a discussion of pleasure rarely enters trade and academic landscape writing. Professional publications often talk of the site, the client, the plant materials, per-haps the particular ecological system or cleverness on the designer's part in solving a particularly thorny drainage problem. More recently, some discussion of the alignment of the garden's axis to the summer solstice or its relation to some geomantic construction might also come into play. The lay publications, in contrast, discuss the delight of the garden, and that making one is so easy—like summer cooking recipes—you can do it in, or to, your own backyard. Color and fragrance and delight are givens; and it is the perfect place for a barbecue. Magazines such as *Sunset* have expanded the world of the house and the garden to the world of lifestyle.

Today might be a good time to once more examine the garden in relation to the senses while putting conscious mental rationalizations on the back burner, to create a mixed metaphor. Although the world's peoples vary greatly in terms of linguistic and cultural matrices, we do share roughly similar human senses, although admittedly these can be honed or dimmed by culture. Is there not a link between the senses and significance, or is meaning necessarily restricted to the rational faculties [6-17]? Barthes would argue that there is a connection. "What is significance?", he writes, "It is meaning, *insofar as it is sensually produced*" [italics in the original].[49]

When an interlocutor once accused Charles Eames of designing furniture only for himself, the designer openly admitted that he did. But he did not design for what was idiosyncratic to himself alone, but as indicative of the greater population of chair users. Why not reinject the same sense of pleasing the individual or self into the landscape design? I do not talk here of Gaia and other forms of touchy-feely expression that constitute yet another form of Neo-Archaicism—since telephone lines have superseded ley lines—but of trying to understand at what level our experience can be shared by others. Not as an abstract symbolic system referring back to Celtic times, but places—and ideas—that acknowledge our time, our sensitivities, and our people. This takes more than a pseudo-significant landscape loaded with the designer's explanatory voice-over, or captions built into the landscape itself. It would seem that a designer could create a landscape of pleasure that in itself would become significant. "Art should not simply speak to the mind *through* the senses," wrote Goethe, "it must also satisfy the senses themselves."[50]

[6-17]
MAN ON THE ROCKS. SAN FRANCISCO. 1975. A primal marking of territory, acknowledging the climate, and creating limited comfort. The fantasy is provided by the reading material, however, not the landscape design.

There are various arguments for a concern for pleasure in garden design.[51] For one—at running the risk of sounding too Californian —pleasure can be a valuable pursuit in itself, as valid as the pursuit of meaning. Even Vitruvius constructed his triad of desirable archi- tectural qualities on commodity, firmness, and delight.[52] In the past, sensory pleasures have served to condition meaning; consider the expression of taste in the selection and arrangement of cut flowers in Japan or the ecstasy of religious experience that underwrote so much Counter-Reformation art and architecture. Sensory experience *moved* the viewer, causing him or her to reflect upon religious meaning as well as one's position in the universe—powerful stuff indeed. Third, despite the influence of culture, individual physiological characteristics, and even transitory psychological state, pleasure is still more predictable than meaning. As in the past—and despite the collapse of collective social norms—pleasure may provide a more defined path toward meaning than the erudite approaches to landscape design discussed earlier in this paper [6-18].

Significance, I believe, is not a designer's construct that benignly accompanies the completion of construction. It is not the product of the maker, but is, instead, created by those who follow: those who occupy, confront, and ultimately interpret. Like a patina, significance is acquired only with time. And like a patina, it emerges only if the conditions are right.

Originally published in *Landscape Journal*, Volume 14, Number 1, Spring 1995.

Acknowledgments

I wish to thank Dorothée Imbert, J.B. Jackson, Karen Madsen, Robert Riley, Simon Swaffield and *Landscape Journal*'s anonymous reader for their perceptive and helpful comments on earlier drafts of this paper. Given the elusive nature of the subject, however, I doubt that I was able to address all their questions and must take responsibility for any shortcomings in both the argument and writing.

Notes

1 For example, the Fall, 1988 issue of this journal [*Landscape Journal*], guest edited by Anne Whiston Spirn and called "Nature, Form, and Meaning," was devoted to just this subject. As might be expected, the range of approaches to the subject was broad, and the resulting interpretations broader still.

2 See D. W Meinig, ed., *The Interpretation of Ordinary Landscapes*. New York: Oxford University Press, 1979. Like anthropologists, cultural geographers read the landscape as a text and are relatively reticent to make judgments about it, much less those of aesthetics. Others, like W.H. Hoskins, however, may decry the modernization of the English landscape and may appraise these residue of culture process based on personal values. See also W.H. Hoskins, *The Making of the English Landscape*, Harmondsworth: Penguin Books, 1955, and compare with J.B. Jackson, *Landscapes*, Amherst, Mass.: University of Massachusetts Press, 1970.

3 Mark Francis and Randolph T. Hester, Jr., eds., *The Meaning of Gardens*, Cambridge, Mass.: MIT Press, 1989. The book, developed from a conference held at the University of California at Davis in 1987, should be distinguished from the previously released typescript and un-illustrated proceedings.

4 Robert B. Riley, "From Sacred Grove to Disney World: The Search for Garden Meaning," *Landscape Journal*, Volume 7, Number 2, Fall, 1988, p. 138. Whether Riley's statement encompasses history as well as contemporary life was not spelled out. The author's hesitation to assign meaning to gardens may stem from the pluralistic and multicultural composition of the contemporary American population. One could add, somewhat desperately perhaps, that this admittedly diverse population appears to be bent on expressing its constitution of differing cultural groups rather than examining the shared characteristics of all human beings.

5 Laurie Olin, "Form, Meaning, and Expression in Landscape Architecture," *Landscape Journal*, Volume 7, Number 2, Fall, 1988, p. 159. The problems that develop from dividing the production of meaning into distinct categories are obvious to the author of the article as well as to its readers. I imagine that Olin would agree that meaning ultimately derives from both categories operating simultaneously.

6 Olin's classifications roughly parallel the two categories I had once proposed in discussing the idea of formalism in the landscape. The first, *trace*, was the unintentional marking or making of space through use. The second, *intent*, concerned conscious spatial definition and/or construction that considered dimensions beyond that of function; that is, the semantic as well as the syntactic aspects of landscape design. Marc Treib, "Traces upon the Land: The Formalistic Landscape," *Architectural Association Quarterly*, Volume 1, Number 4, 1979 [included in this volume].

7 The library on meaning in philosophy is vast and makes trying reading. The award for the most provocative title should probably be given to Cambridge

dons C.K. Ogden and I.A. Richards, *The Meaning of Meaning*, New York: Harcourt, Brace & World, Inc., 1923. If one accepted their definition of meaning in *language* and extended it to *landscape architecture*, one would have to agree that it was indeed possible to design meaning into landscapes: "The meaning of any sentence is what the speaker intends to be understood from it by the listener" (p. 193). The authors, obviously, make no such claim for landscape design, however, nor do they infer that linguistic theory is applicable in any form to the making of landscape.

8 For example, I am told that—like landscape meaning—there remains no clear definition of electricity. This has not hampered our ability to understand, produce, modulate, and utilize the resource, however. Louis Armstrong is said to have said that if he had to explain jazz to someone, they would never really understand it. This just may be true of meaning as well.

9 That this stance is problematic seems obvious.

10 Tunnard's essays that would constitute his 1938 *Gardens in the Modern Landscape* appeared serially in *Architectural Review* starting the previous year. About the same time, James Rose contributed a series of articles to *Pencil Points*, the predecessor of today's *Progressive Architecture*, including one essay entitled "Plants Dictate Garden Form," written in 1938. The conclusions that Rose reached in this essay closely paralleled those of Tunnard. Both included a list of plant materials suitable for modern conditions.

Alas, Progressive Architecture *has gone the way of* Pencil Points *since this essay was published.*

11 Conversation with the author, June 1988; but he still says it.

12 The notable exception was the work of Roberto Burle Marx, who would frequently be characterized as a "painter in plants," and who drew on the shapes of modern, often non-objective, art. See Marc Treib, "Axioms for a Modern Landscape Architecture," in Marc Treib, ed., *Modern Landscape Architecture: A Critical Review*, Cambridge, Mass.: MIT Press, 1993.

In a series of articles published in *Architectural Record* at the beginning of the 1940s, Rose, Eckbo, and Dan Kiley linked the physical and social environment from the intimate to regional scales as a prerequisite of responsible landscape architecture. They did not talk of significance, however, but implied that meaning accompanies an intelligent design, or that it was just not an issue. These articles have been republished in Treib, *Modern Landscape Architecture*, cited above.

13 In Gideon's eyes, space was the primary quest of modern architecture, the realization of an adventure he traced back to Baroque spatial planning in Rome under Sixtus V and the undulating facades of Francesco Borromini. Sigfried Gideon, *Space, Time and Architecture*, Cambridge, Mass.: Harvard University Press, 1938. The quest reached an apogee in Bruno Zevi's *Architecture as Space*, New York: Horizon Press, 1957.

14 For a representative collection of Greenberg's ideas and writings, see Clement Greenberg, *Art and Culture: Critical Essays*, Boston: Beacon Press, 1961.

15 Garrett Eckbo, *Landscape for Living*, New York: Reinhold Publishing, 1950. See also Reuben Rainey, "'Organic Form in the Humanized Landscape': Garrett Eckbo's *Landscape for Living*," in Treib, *Modern Landscape Architecture*, pp. 180–205.

16 Ian McHarg, *Design with Nature*, Garden City, NY: Doubleday, 1969. Despite his predominant polemic and pervasive rationale, McHarg admits moments of poetry and suggestions of meaning: "The best symbol of peace might better be the garden than the dove" (p. 5).

17 Olin, "Form, Meaning and Expression," pp. 150–151.

18 To my mind, one of the real burdens of landscape architecture is that two professions are combined under the same name, as if their interests and goals were coincident. Landscape architecture is concerned with forming, as well as planning, a landscape; landscape management, or planning, with its regulation. Obviously, they overlap in their concern with living systems, but landscape architecture requires active formal intervention in a way that regional planning does not. This is not to say, however, they both do not have consequences in the form of the landscape.

19 In his 1984 *California Scenario* plaza/garden in Costa Mesa, California, Noguchi appears to have adapted a wedge-shaped fragment of the astronomical observatory for use as a water source. While this element can also be read as a modernist abstraction of a hill, the form bears a striking resemblance to its Indian predecessor. Noguchi's program for the garden encompassed the various ecological zones of California, from mountain meadow to desert: an attempt at evoking the genius loci and creating meaning?

20 Gary Dwyer, "The Power under Our Feet," *Landscape Architecture*, Volume 76, Number 3, May–June, 1986, pp. 65–68. The choice of Ogham as the script with which to inscribe the fault line was based on its formal properties alone: it is written as cross marks across a linear spine. Dwyer himself asks the critical question: "How can an ancient Celtic language have anything to do with the San Andreas Fault?" And replies: "Aside from its development by a primitive people who were rhythmically allied with the forces of nature, Ogham began like all languages with the mark, with 'naming the unknowable.'" The substantiation remains unconvincing to me.

21 After an exhaustive search, and a telephone call to the author, I have been unable to find the exact source of the quotation, or even whether this was the exact quotation. If not precisely those words, the spirit of Professor Howett's observation is captured by them.

22 Christian Norberg Schulz determined three ways in which man-made places relate to nature. The first regards rendering the natural structure "more precise"; in the second, construction complements the natural order, while the third symbolizes it: "The purpose of symbolization is to free the meaning from the immediate situation, whereby it becomes a 'cultural object,' which may form part of a more complex situation, or be moved to another place," *Genius Loci*, New York: Rizzoli, 1980, p. 17. Edward Relph's *Place and Placelessness* constitutes, in some ways, the complement to Norberg Schulz's more naturally oriented vision. Relph also includes the symbolic as part of the triad of factors that create the sense of place: "The identity of a place is comprised of three interrelated components, each irreducible to the other—physical features or appearance, observable activities and functions, and meanings or symbols." London: Pion Limited, 1976, p. 61.

23 My anecdote is a paraphrase of a citizen reaction I overheard in 1990 when photographing an urban triangle in Washington, D.C., planted by Oehme/Van Sweden. My kibitzer read my taking of photographs as documenting a deplorable urban condition, pre-

sumably as evidence to have the wrong righted as soon as possible. This was in spite of the fact that numerous plaques identified the various grasses, and clueing at a second level, that the wild look was intentional.

24 I realize, of course, that there are far more considerations bearing on these decisions than the aesthetic alone. But at some point in the process, aesthetic questions must be addressed.

25 "If one wishes to work on the cutting edge in either fine art or design," writes Martha Schwartz, "one must be informed of developments in the world of painting and sculpture. Ideas surface more quickly in painting and sculpture than in architecture or landscape architecture, due to many factors including the immediacy of the media and the relative low investment of money required to explore an idea." "Landscape and Common Culture," in Treib, *Modern Landscape Architecture*, p. 264. As is often the case, however, by the time art ideas are applied to landscape design, they are a bit tired and worn. For a passionate argument for the Neo-Archaic in art—one source of landscape architecture in the 1980s—see Lucy Lippard, *Overlays*, New York: E.P. Dutton & Co, 1983.

26 Norma Evenson, *Paris: A Century of Change*, New Haven, CT: Yale University Press, 1978. Or taken at rush hours, as a loop and linear parking lot.

27 In fact, one of the slaughter houses, La Halle aux Bœufs, was renovated into an art space by Reichen and Robert in 1985; a modern recent structure for animal dispatch was heroically recast as the City of Science and Industry by Adrian Fainsilber in 1987 and is the park's principal attraction.

28 The notable exception is Alexandre Chemetoff's *Sequence IV*, or Bamboo Garden. Given its

sense of path, its enclosure, and its Didactic revelation of subterranean services, the bamboo garden is both a lesson and a respite from both the city and the other parts of the park at La Villette. As of June, 1993, however, it had become overgrown and is in need of pruning and reformation.

29 Much of what has been written perpetuates the designer's original claims; many of the authors seem never to have visited the actual park and their writings are discourse about discourse.

30 According to J.B. Jackson, expediency is a hallmark of vernacular building. *Discovering the Vernacular Landscape*, New Haven, CT: Yale University Press, 1984.

31 This friend, who happens to be French and writes about modernist French gardens, wishes to remain anonymous.

32 *Since this essay first appeared, the movement has gained greater visibility under the name "eco-revelatory design." See* Landscape Journal, *Special Issue, "Eco-Revelatory Design: Nature Constructued/ Nature Revealed," 1998.*

33 Garreau offers two "laws" that govern the naming of developments. First, there is Jake Page's Law of Severed Continuity: "You name a place for what is no longer there as a result of your actions." Next, "The Keith Severin Corollary": "All subdivisions are named after whatever species are first driven out by the construction. E.g., Quail Trail Estates." In "The Laws: How We Live," *Edge Cities*, Garden City, NY: Doubleday, 1991, pp. 461–471.

34 Sir Geoffrey Jellicoe's proposal for the Moody Gardens in Galveston was essentially a landscape theme park, evoking (but not copying) historical garden types. Jellicoe, *Landscapes of Civilization*, Woodbridge, Suffolk: Garden Art Press, 1989.

35 There is some indication that they do. Particularly in good weather, the less formal areas of the "gardens in movement" are highly utilized, perhaps because they provide some of the only truly private—and shaded—spaces in what is otherwise a highly structured ensemble. In that way, they resemble the country in comparison to the city.

More than ten years have passed since the Jardin en mouvement *was planted, and over the years the variety of species has declined as the shrubs have matured. For before and after photos of the garden, see "The Content of Landscape Form [The Limits of Formalism]," the following essay in this volume.*

36 I realize that there is quite another school of thought that regards both human experience and significance as more or less universal. This belief has produced "pattern languages," among other theories, derived from a selective potpourri of peoples and places, with the assumption that the proper blend (selected and structured by the authors) will perfectly suit all of humanity—certainly, at least, twentieth-century America. My own experience through travel and reading—supported by historical study—suggests quite the opposite; that is, that values are not universal, but instead are particular to a people, place, and time. Perhaps this could be appropriately termed "cultural relativism"—and it probably has been so termed by someone somewhere.

37 Thus, Japanese gardens built outside Japan are mere shadows of their referents since they lack their native cultural matrix. They become "japanesque" and expose physical features as a photograph captures an image but only rarely the essence of subject.

38 Polite, like the term Monumental or High Style, is used in this essay in (near) opposition to the Vernacular tradition of

landscape making and building. It implies neither a rank ordering of one above the other nor any particular character—except that the Polite tradition will normally approach environmental design far more self-consciously than the Vernacular.

39 "It is doubtless a difficult notion to appreciate today, but in the eighteenth century all the fine arts were deemed to have representation at their center, and gardening aspired to *beaux-arts* status," John Dixon Hunt, "The Garden as Cultural Object," in Howard Adams and Stewart Wrede, eds., *Denatured Visions*, NY: Museum of Modern Art, 1991, p. 26.

40 Ibid, p. 28.

41 Ibid.

42 In his or her notes, the reader anonymously reviewing a draft of this essay for *Landscape Journal* wisely noted two categories of meaning: "A. Systems of Signification/Representation in the landscape (metaphysical, narrative, allegorical, symbolic), and B. Circumstances of engagement with the landscape (experiential, sensory, physical)." This might be interpreted broadly as meaning that accrues perceptually as opposed to meaning that accrues conceptually.

43 It would be interesting to return to the Viet Nam memorial in a hundred years' time to determine whether the design and the inscribed names would retain their effect.

44 Folk cultures have been described as those that are geographically delimited, developing only slowly over time. Mass culture, in contrast, is more broadly ranged and changes rapidly.

45 Or as Robert Riley put it: "Such a lack of shared symbolism does not rule out the garden as a carrier of powerful meaning but it does discount the likelihood of meanings that speak strongly to the whole society." Riley, "From Sacred Grove to Disney World," p. 142.

46 J.B. Jackson, among others, has pointed out that the ocular-centric garden is a Renaissance development and that during the medieval and earlier ages the correspondences between plant and cosmos were firmly established. The form of the plant or its fragrance or its name suggested its value through associations. A yellow plant might be appropriate for curing jaundice; a round one might assuage headaches. Those that cared about such things—admittedly, a small community—were bound together in a common belief system through Christianity. "Gardens to Decipher and Gardens to Admire," in *The Necessity for Ruins*, Amherst, Mass.: University of Massachusetts Press, 1980, pp. 37–54.

47 Robert Riley cited Mary Douglas's term "condensed symbols" that "carry not just one meaning but accretions of many meanings, layered upon each other and over time. They are symbols that are commonly agreed upon, not designer-chosen, that connote deep affective meaning, not quick cleverness, and that are integral to a context that is culturally agreed upon as appropriate." Riley, "From Sacred Grove to Disney World," p. 142.

48 "Does our involvement with publications enter here? While neither meaning nor pleasure can be photographed, there can be pleasure depicted within a photograph; the photograph itself can provide pleasure, of course." Roland Barthes, *The Pleasure of the Text*, NY: Hill and Wang, 1975, p. 6.

49 Ibid.

50 J.W. von Goethe, "The Collector and His Circle," *Propyläen*, 1799, in John Gage, *Goethe on Art*, Berkeley, CA: University of California Press, 1980, p. 70.

51 My own thoughts on this subject have been greatly augmented by suggestions from Robert Riley, for which I am grateful.

52 Vitruvius, of course, spoke Latin, not English. This particular rendering of the Latin original is by Henry Wotten.

7

The Content of Landscape Form
[The Limits of Formalism]

2001

I.

How do we evaluate and appreciate landscape architecture? Is it the skill with which the walls, rills, and floors have been designed and crafted, the power of the spaces—the formal beauty alone? Or do we praise the success with which the spaces please us, how they provide warmth in a cold climate, a sweet fragrance among dust, or places for sitting and human conduct, or settings to eat or to dream? Do we appreciate a design because it seems so appropriate to the climate or to the topography, or as an escape from it? Do we reward the landscape for using a minuscule amount of water in a desert landscape, no matter the corollary sensual deprivation?

The question of appreciation and evaluation informs the greater question of landscape content. Of what value is a landscape design; what is its content? It has been said that since there is no landscape without content, so can there be no work of landscape architecture without content. This assumption has particular resonance if one believes, as I do, that meaning derives from the transaction between the perceiver and the artifact.[1] According to this way of thinking, the designed landscape serves essentially as the material for sensing and interpretation. Ultimately, comprehension and pleasure rest with the individual influenced by his or her cultural matrix. Of course, other schools of thought do exist, and several of them hold that it is possible to imbue meaning in the course of design and making, especially in cultures bound within a common system of belief. In this essay, I would request a temporary suspension of disbelief from those who follow this latter view. Here I would propose that the content of landscape architecture is the raw material to be transformed through design, material from which we may derive pleasure and/or significance. What sort of raw material, its potential and its relevance, is the essay's base subject. Of the panoply of possible sources for content, for reasons hopefully explained below, I will focus on ecology, social/historical aspects, and form (and space) themselves.

II.

In recent years, that is at least since the mid-1980s, landscapes structured by patterning, realized in synthetic materials and restricted in vegetation have received considerable attention and widespread publication. For the most part these designs make vivid retinal images and striking photographs; they are experienced on site as exercises in order and form. They may be beautiful or ungainly, pure or assembled, uniquely crafted or drawn from varied vernacular and industrial elements. While their visual interest is, for the most part, undeniable, experienced as landscape—considering the full potential offered by a designed landscape—they remain circumscribed and limited. What works in the photograph does not necessarily thrill on site or maintain continued interest over time. Is this because the work itself lacks sufficient interest, or that the photograph (through isolation, recomposition, idealized lighting conditions, etc.) has so increased the power of the place that it is difficult to match in actuality?[2]

Since its invention in the nineteenth century, the photographic image as printed, or more recently digitized, has exerted a potent influence upon the formulation and witnessing of the designed landscape.[3] Of course, photographic experience is by its very nature more narrowly limited to the visual sense, in turn, suppressing the haptic, olfactory, auditory, and temporal dimensions of landscape perception. The result —sadly, to my mind—reduces the potentially manifold dimensions of experience to only two. In the process, the formal aspects become the purpose or content of the design; the image reigns supreme.[4]

While the skills of design, construction, and detail certainly constitute content in and of themselves, there are limits to the continued effect of this formalism and the attitude with which it regards the environment and society. Form as content is an old story in modern painting, of course, and to a lesser extent in architecture. Critic Clement Greenberg argued that painting, before it ever represented any subject external to its physical dimensions, was essentially a question of marks made upon a field. By extension, the paintings that most clearly manifest that definition—those free of the burden of mimesis—should be more highly regarded. This led to a quest for "flatness," which remained a central concern of painting for decades after World War II.[5]

Rather than structure, space, and pattern as content, deeper works may result from using these as vehicles for embodying other types of substance, among them the understanding and judicious application of ecological processes (including those of the immediate as well as the greater site over time), and the regard for humans singly and in groups, contemporary and over time. The manner in which the designer addresses these factors may also elevate a physical statement of these concerns, alone or together, to a poetic level. It is admittedly a difficult task, and without question, no work is ever perfect in all respects. Nonetheless, several landscape architects currently in practice have undertaken designs with these considerations at their core. As examples I cite several projects by Hargreaves Associates in the United States, and Georges Descombes and Dieter Kienast in Switzerland.

III.

In the last few decades, the pendulum that traces the evolution of design styles has once again swung regularly between architectonic and more naturalistic manners. Perhaps this comes as no surprise. Since the beginning of landscape architecture as a defined practice, the manner in which we construct landscapes has displayed these alternating modes, with an almost complete gradient of variations in between the two extremes.[6] The Garden of Eden is normally conceived as a natural landscape, for example, albeit bounded by an excluding wall. And from their very origins, environments managed for agricultural production have required a more efficient organization of planting and irrigation, leading to landscapes in which the human hand has been more, rather than less, apparent.

Although convenient for historical studies, this two-part division into formal and informal is only partially useful. For one, it favors the sense of sight to a great degree, undervaluing the significance of botanical or cognitive processes. The power of garden design as visually perceived, it would seem, instead rests on the overriding scheme and the balance of the garden's elements, and the collisions and transitions among them. There is no formality or informality in isolation, as there is no concept of nature free from a concept of culture. Never is the question of formal and informal one of simple opposition, or a simple choice of one over the other. Most importantly, we need to question to what extent the forms, the space, and the manners of realizing landscape design truly embody its content.

While there are many problems regarding the merit of landscapes only in terms of form and space, even the most logocentric critic must admit that it is just these aspects that ultimately confront human perception. As such, they seem virtually impossible to avoid.[7] It is useful to further comprehend the reasons behind formal manufacture, and here art critic Dave Hickey's discussion of the painter Bridget Riley is instructive. Hickey distinguishes between perceptually and cognitively intended art works, further dividing the more formally instigated category into two groups of varying value. He discusses three categories in this way:

> [T]he rhetorical-empirical brand of "behaviorist modernism" practiced by Bruce Nauman and Richard Serra, for whom, as for Riley, the manipulation of material and formal means is directed toward the evocation of a local, cognitive-kinesthetic experience that is quite distinct from linguistic communication (which presumes that the work of art bears a message) and formal appreciation (which posits the work of art as a dead thing, artfully manipulated and sensitively perceived).[8]

Do formally conceived landscapes serve greater purpose—"local, cognitive-kinesthetic experience," for example—or do they exist only for "formal appreciation (which posits the work of art as a dead thing …)?" Most paintings and sculptures are more finite than a designed landscape, no matter how perfectly maintained. If perception is the primary vehicle for understanding, we also need consider aspects of cognition that are equally, if not more, crucial for maintaining interest and pleasure—and for evaluating landscape merit. This mental discernment distinguishes between the manner of *making* a landscape and how we *think* about that landscape. It again raises the issue of landscape content.

Thus, we might gauge the content of landscape design along three axes: the formal (which includes space, form, and materials); the cultural (which includes history, social mores, and behavior); and the environmental (among them ecology, topography, hydrology, horticulture, and natural process). Of these—and I admit here to personal bias—the formal serves best as a means to an end rather than an end in itself.

The American cultural landscape historian John Brinckerhoff Jackson defined landscape as "a space on the surface of the earth; intuitively we know that it is a space with a degree of permanence, with its own distinct character, either topographical or cultural, and above all a space shared by a group of people."[9] This definition suggests that basic to all landscapes—whether designed for functional, contemplative, or entertainment purposes—is the presence and accommodation of human beings as individuals or in society, serving their physiological or psychological needs.

In addition, the conditions particular to the location also inform the making of the landscape—although I would not go so far as to say they truly *determine* an approach. Thus, landscape design—consciously or not—always reflects contemporary values and attitudes; there is no single way to create a landscape, even at any particular time. Making places in an arid zone, for example, could follow several paths. The designer could accept the limitations of the desert and frame existing topography and vegetation, as did Frank Lloyd Wright in 1938 at his own home and studio Taliesin West in Arizona [7-1]. Or the desert could be approached more abstractly, as non-professional garden-makers often do [7-2] in suburban Phoenix.[10]

But one needn't accept the limits imposed by local conditions: a landscape could also be conceived as a vehicle to transcend the demands of everyday life. The idea behind the paradise garden, for example, has always been to escape what nature offers, to develop irrigation systems and horticultural methods that would allow landscapes to deny strictures of local conditions. In the minds of many peoples, paradise proposes the antithesis of where a people live, as Yi-Fu Tuan has shown in his book *Topophilia*.[11] To those living in cold climates, paradise is warm and lush; to desert folk, it is lush and well-watered.

The exquisite Patio of the Oranges in Seville embodies the paradisiacal idea, where the golden fruit and the enjoyment of shade derive from an adroit management of irrigation [7-3]. At the Patio of the Oranges the technique—the formal organization, the details, the true design —is more obvious than any evocations of paradise based on a more naturalistic model. Without doubt, we do read and appreciate this garden, like many flower gardens, on formal terms alone. But this initial pleasure may be heightened by appreciating this landscape across more than one dimension. Thus, as landscape architects or artists or architects we may appreciate more rapidly the beauty with which this religious courtyard has been made. This produces somewhat of a dilemma, with questions such as those that opened this essay. How do we weigh the value of a designed landscape?

As we probably cannot accept any simple opposition of formal and informal to categorize landscape form, so can we not evaluate landscapes using any one of these three classes of considerations taken in isolation. Should we not evaluate landscapes using all three sets of issues: cultural, environmental, and formal (here "formal" describes the properties of form and space rather than style)? Social accommo-

dation without a consideration of the place may lead to uncomfortable landscapes. Surrender to the restrictions of climate may yield landscapes devoid of beauty or human appeal. Visual beauty alone risks the danger of being sterile and removed from life. Engaging the entire trio to both create and evaluate landscape architecture appears to offer far greater return. Cultural concerns, often translated into planning for use, need to be taken quite broadly. By Western standards, gardens in Japan such as those created for the Zen sects appear to have no function [7-4]. Yet contemplation, dreaming, and aesthetic appreciation at times are all valid landscape functions in and of themselves. On those grounds, the dry gardens perform handsomely as cultural vehicles.

With this proposition of values, the focus now shifts to several selected tendencies in recent landscape design practice. My principal concern here is the escalating appreciation of landscape design via the photograph or cinematic image, and more recently, as digital representation. That we now more often look at representations rather than actual landscapes has allowed formalist designs to achieve great prominence. Other aspects of the landscape, more subtle or less easily conveyed in photographs and publications, have suffered neglect.[12] As a result, we often reward form-as-content (which, as noted above, it can be), rather than form and space as what the painter Ben Shahn once termed "the shape of content."[13]

Discussing content questions the medium by which most landscapes are known today: the photograph in publication. In many cases, it is the visual appeal of the landscape—or even the appeal of the photograph alone—that seduces the viewer. There may be no appreciation for the managing of the constraints that guided the design and coerced the true brilliance of its solution. Since viewers of the photograph rarely attend the actual landscape, the experience of the photograph substitutes for the experience of the place. As a result, we "filet" the content by appreciating only the look of the design.

In some ways this may not be a completely negative practice, as even in photographs new ideas enter the landscape discipline and practice. On the other hand, engaging images of landscapes by Peter Walker or West 8 or Martha Schwartz are copied in almost every country on Earth, with little regard for their possible ill fit within an alien situation [7-5]. But even here, in this worst case scenario, some latitude must be granted. If the landscape architect appropriating these forms understands the specific conditions of his or her own society and environment,

[7-2]
GRAVEL GARDEN.
PHOENIX, ARIZONA.
1998.

[7-3]
PATIO OF THE
ORANGES.
SEVILLE, SPAIN.
SIXTEENTH
CENTURY+.

[7-4]
DAICHI-JI.
SHIGA, JAPAN.
SEVENTEENTH
CENTURY.

perhaps design does become principally a question of formal idiom. Perhaps. The danger of blind copying, however, is that it tends to replicate patterns and forms without any real consideration of the local conditions or their consequences. Perhaps a grid of squares or a diagonal line of planting beds just doesn't make much sense in the Swedish forest. Too often, we are skillful at copying forms as portrayed in photographs without investigating to sufficient depth the ideas behind them.

How can we balance these factors? How do we acknowledge contemporary needs and contemporary programs? How can we interpret lessons from histories, both local and exotic? How do we address ideas of contemporary culture or related art forms and thinking in other disciplines? Only those most conservative would argue that landscape architecture should not advance with its culture and with its times. There still is merit in the modernist belief that only rarely does a historical answer serve us as a precise model for contemporary life, although history does aid our understanding of the present and the future.

A highly selective sample of work from the very recent past may help answer some of these questions.

IV.

Partly in reaction to the then-prevailing analytical and usually naturalistic manner supported by the writings and practice of Ian McHarg, a group of landscape architects in the 1980s and into the 1990s came to rely to a large degree on formal pattern. Peter Walker in the United States and later West 8 in the Netherlands (and followers worldwide) have utilized stripes, grids, rotated geometries, and regularity to structure their design.[14] Even when addressing ecological requirements, these landscape designs are in many respects a return to the parterres characteristic of the seventeenth-century French garden—except that now the parterre and the garden as a whole are rendered congruent.

A number of these works, alas, are more stimulating in photographs than in reality. The photograph superimposes a rectangular frame upon the landscape, against which are composed the linear rotations or regular bed plantings.[15] The photograph is a fragment that is forced to represent the whole, like the literary trope, synecdoche. But a landscape is not a fragment: it *is* a whole, and at times these designs maintain our interest only at small scale for short periods of time. At Burnett Park in Fort Worth, Texas, for example, the overall pattern is arresting in its overlays of orthogonal and diagonal lines, and their relation to the structure of the park as a whole [7-6].[16] The pattern, which in this case was said to derive from pedestrian movement across the park, on first look, appears to be a dynamic of oblique lines. Yet in the end, the net experiential effect is quite static, and little draws us from point A to point B because point A is just about the *same* as point B. In this sense, the more extreme examples of formal patterning —espccially those that remain flat, without true spatial consequences —demonstrate little regard for the human body, mystery, and appeal, or for senses other than vision.

The use of pattern has had success, however, and is not so easily dismissed. Two examples of landscapes from the mid-1980s stand as the high points of recent formalism. The first comprises the landscapes designed for the IBM community at Solana, outside Dallas, Texas; the Office of Peter Walker Martha Schwartz were the landscape architects [7-7]. The architect for the Solana IBM campus and the town center was Ricardo Legorreta; Mitchell Giurgola designed the West Campus. The power of the IBM scheme derives from the graphic structure's directing spatial development rather than remaining a two-dimensional figure alone. In the main office complex, for example, the architect convoluted the perimeter of the building blocks, using courtyards as transition spaces between the landscape of the architecture and the architecture of the landscape. These courts function as hinges that pivot the eye and the body from inside to outside and vice versa. Zones of varied shades and enclosure result, providing comfort during those times conducive to occupation outdoors, and visual pleasure through-out the year [7-8]. Legorreta's vibrant color palette intensifies the architectural presence and heightens the visual articulation by setting yellow and purple planes against the greens of the vegetation and the blues of the canals.

This is the key: at Solana, pattern instigates and structures its three-dimensional consequences and in the process becomes spatial; architecture and landscape architecture enfold within a charged equilibrium, geometrically conceived, formal in its gratification. However, this aesthetic pleasure—as noted above—also derives from modulating climate and light for physical comfort. Success is measured along more than the aesthetic axis alone, even if it may not have been the designer's primary concern.

The 1984 plaza/terrace for the North Carolina National Bank (NCNB) in Tampa, Florida, is more rigorous than the Solana landscapes in using purely geometric structure [7-9, 7-10].[17] Here Dan Kiley superimposed several layers of gridded pattern to generate a complex chessboard for trees and ground covers rather than for bishops, rooks, and castles. The predominant grid established a field of Washingtonia palm trees upon the ground plane, which is in fact the roof of a parking garage below. Against this orthogonal organization sweeps of clumped plantings of crape myrtle bloom a vivid pink in springtime. The trees appear irregularly spaced but in fact follow judicious placement on a smaller-scaled grid. The ground plane demonstrates, by far, the most complex patterning, composed by alternating strips of paving and zoysia grass. In the design of the office tower that the terrace serves, architect Harry Wolf utilized the ratios of the Fibonacci numbers to modulate the proportions of the architecture. Kiley extended this mathematical thinking into the garden, using these progressions to generate the varying balance of grass to paving from the bank building to the far extent of the terrace. This admittedly highly, almost maniacally, structured effect produces a complex, though equilibrated, composition whose readings continually change with the visitor's position.

Water creates the garden's fourth layer. Glass roofs the entry corridor to the parking garage. In its early years, the water that filled this roof functioned as an irrigation canal that fed a series of rills and fountains that penetrated deeply into the greenery of the garden. There is little doubt that the design drew inspiration from the garden tradition of Moorish Spain, and it is difficult not to recall the magnificent gardens and courtyards of the Alhambra while strolling in Tampa. Yet despite these historic references, through Kiley's rigorous geometries and masterful play of linear elements against those more massive—and more importantly perhaps, the development in space of the four layers of the vegetation and water—a garden perceptually rich has resulted. There is little doubt of its contemporaneity, and yet there is little that does not suggest some historical, classical precedent. This is the magic of the Kiley manner, and it demonstrates that the past always maintains its relevance to the present as a source of learning through discerning transformation.

Despite the appreciation and enjoyment of landscapes such as these, problems do result from using pattern-making as the basis for landscape design. For example, varied orientations or slopes require differing planting solutions—and yet the continuity of the pattern

[7-5]
(above left)
VSB BANK GARDEN.
UTRECHT,
NETHERLANDS. 1995.
WEST 8.

[7-6]
(above right)
BURNETT PARK,.
FORT WORTH, TEXAS.
1983.
SWA / PETER WALKER.

[7-7]
(middle left)
IBM SOUTHLAKE.
SOLANA, TEXAS. 1989.
OFFICE OF PETER
WALKER MARTHA
SCHWARTZ.

[7-8]
(middle right)
IBM SOUTHLAKE.

[7-9].
(below left)
NCNB BANK TERRACE.
TAMPA, FLORIDA.
1988. DAN KILEY.

[7-10]
(below right)
NCNB BANK TERRACE.
The swaths of crape
myrtles.

demands a repetition of similar elements. At the powerful walled entrance gateway to the Solana complex, for example, the four slopes each face different directions, and the tops of the hills have different drainage conditions than their bottoms [7-11]. As a result, it has proven difficult to maintain a matching pattern of the striping on four slopes.[18] Perhaps greater consideration of these factors would have modified —and conceivably, enriched—the pattern; or further study may have determined that arrangements of rocks or gravel would have been a better way to execute the stripes. The two manners—formal and eco- logical—are not antagonistic, unless the first is employed with little regard for the second. Several horticulturalists have noted the ever-present danger of planting trees of a single species. If one should fall ill, all its neighbors may be tainted and threatened. And if too many die in one area, the pattern is destroyed. This, too, has remained a constant threat to the Solana landscape.

Despite these cautions, however, the success of the Solana and Tampa projects demonstrates the sizable potentials for these architectural landscapes, particularly if provided the means to maintain them. They emphatically remind us that the formal tradition will not disappear, and that it can acquire renewed vigor in contemporary times by drawing upon influences such as minimal art, mathematical progressions, and even historical reference. Any problems concerning the selection of tree species, one would believe, can be solved through shrewd selection. More critical remains the continued need for focusing on geometric structuring that unfolds as truly three-dimensional and spatial rather than as pattern-making that begins—and ends—as a flat surface.

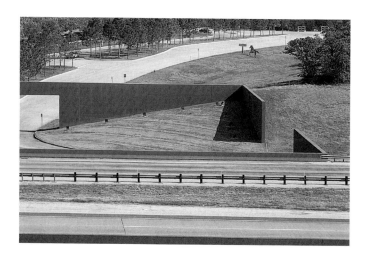

[7-11]
IBM, SOLANA, TEXAS.
1989. OFFICE OF
PETER WALKER
MARTHA SCHWARTZ.
Freeway intersection.

The lack of concern for habitable spaces raises issues about intention and content in landscape architecture. As proposed earlier, could we not agree that human occupation and use are the content of landscape design, and that nature and ecological process constitute the matrix in which we create these new terrains? Landscape architecture thus becomes the compounding of these two aspects into a legible cultural expression, yes, using the formal means we call style.

V.

The work of Hargreaves Associates, based in San Francisco and Cambridge, Massachusetts, exemplifies a heightened interest in form developed from natural process and human use, especially in their designs for a series of waterfront parks. The land for Byxbee Park, located south of San Francisco on the western shore of the bay, comprised garbage and earth fill, in some places measuring more than fifty feet in depth [7-12]. The site, which was intended to become much-needed recreational land, was thus the product of human hands and built on human waste. The governmental sponsor for the park's hundred-odd acres stringently restricted modeling of the earthen contour. A yard-deep cap of soil and clay stabilized the putrefying garbage below, with a flame perpetually exhausting the methane collecting beneath the ground. Because it was believed that any rupture in the earthen topping might allow the escape of noxious gas, no trees were planted.[19] And because seepage might percolate pollutants into the water table below, irrigation was precluded. These constraints directed the

[7-12]
BYXBEE PARK.
EAST PALO ALTO,
CALIFORNIA. 1992.
HARGREAVES
ASSOCIATES.

designers' attention to land contour as the principal design feature and fostered a respect for native species of grasses, completely dependent on rainfall for their nourishment; they were allowed to turn brown during the dry months of the northern California summer.

Despite the troublesome fragmented disposition of the park's design features, as a complete entity the Byxbee project demonstrates that ecology—and entropy—are not antagonistic to landscape design; quite the contrary, an understanding of environmental forces can stimulate significant innovation. The land artist Robert Smithson called our attention to the aesthetic potential of entropic process as early as the 1970s; the effect of entropy on landscape over time was an important aspect of his thinking. But only rarely have its possibilities informed the design of landscapes rather than the making of art.[20]

During the last decade, Hargreaves Associates has designed water-front landscapes in San Jose and San Francisco, California, Portland, Oregon, Louisville, Kentucky, and Lisbon, Portugal. Although each design rigorously investigated precise local conditions, as a group the parks reflect a common attitude toward the processes and meeting of land and water, reforming them in accord with ecological, social, and aesthetic parameters. They also constitute some of the more provocative recent landscape architecture at a larger scale, and in their distinctive approach, they resolve conflicting attitudes within the profession.

The Guadalupe River Park, whose master plan dates from 1988 to about 2002, was intended to reveal San Jose's obscured riverine open space [7-13]. Mission San José had been established in the late eighteenth century as the Spanish established their hold on Alta California. Into this century, the river has remained a green space within the city, but new highways, shifting demographic patterns, and limited access have all impeded its enjoyment. During the boom years of the 1980s — heightened by the growth of the nearby Silicon Valley—San Jose witnessed the active infusion of interest and capital into its downtown infrastructure. The city landscaped primary arteries and constructed major public works including a sports arena, several museums, and the Hargreaves Associates-designed Plaza Park, completed in 1989.

Addressing the threat of inundation and major devastation to the downtown area, the Guadalupe River was itself slated for overhaul by the Army Corps of Engineers. The narrow sliver of a river had frequently flooded its banks—and there was every indication that the increased density of recent construction would only escalate the impact of the next major deluge. Why not use the necessary flood control work to create a three-mile-long park that would bring the people and city to the river and vice versa?

Working with a small army of specialists, the landscape architects developed the general plan and specific designs for the length of the river in the downtown, dealing equally with the "underlay" and "overlay" of the landscape along the banks. George Hargreaves uses the term the "underlay" to describe the responses to flood control during times of heavy winter rains or spring run-off; "overlay," in turn, comprises the more visually apparent aspects of the design: the reforming of the earth and the planting of vegetation. Ordinarily much of this

[7-13]
GUADALUPE RIVER
PARK. SAN JOSE,
CALIFORNIA.
1988–1996.
HARGREAVES
ASSOCIATES.

project would be regarded as civil engineering, but in this instance, the landscape architects took an active interest in hydrology and its consequences in form. Their thinking—developed in close coordination with the Army Corps of Engineers—derived from an understanding of water flow, in a sense abstracting and enlarging the consequences commonly formed along the river's banks.

The design models the land in accordance with the necessities of flood control, settings for use, and the need for plantings along the long linear strips of river. The park's design could be described as a series of flows and interruptions, not unlike a river in itself. The prevalent terrain parallels the axis of the river, but at key points it is reformed into level planes, slopes, and mounds that parallel the patterning and branching that results from the dendritic process.[21] Both the citizens and the profession have acknowledged the success of the design, although additional phases have been left unimplemented—a common fate for public projects of this magnitude.

The winning park competition entry for the Louisville Waterfront Park addressed the site's despoliation from prior use [7-14, 7-15]. Industry had lined the bank for over a century, removing the land from public access, despite the long tradition of public parks in the city. In the minds of the citizens, the land was an absence, seen only from a speeding car on the elevated highway.[22] The landscape architects rejected the idea of bringing the downtown to the river, and instead convinced the city to bring the river to the downtown.

Unlike the San Jose park, the Louisville scheme occupies a single shoreline. The land, given its history of industrial use, required detoxification and regrading; severed from the urban fabric by freeways and bridge approaches, the design proposed replanting trees as links between the park and the downtown. While the scheme in plan appears linear, the park is actually conceived as a chain of event spaces that vary in their function and form from those more open, ceremonial, or activity-oriented, to those more natural and solitary.[23]

From these more urban functions, the park extends northeastward to more private areas, more "agrarian" in appearance and more intimate in scale, until beyond a gigantic bridge the children's park terminates the sequence. As conceived, the Louisville Waterfront Park would be a green park, but not a green park in the Olmstedian mold. The central green, with its play of skewed rectangles and watercourse, may function at times as a community commons for meetings and concerts; on a daily basis it offers broad open spaces for sports, taking the sun, or even flying a kite. In other areas, mixed plantings of conifer and deciduous trees provide change throughout the year yet guarantee spatial closure at all times. A hierarchy of paths supports a variety of movement; again the shaped mounds articulate spaces within spaces that offer a retreat from the wind in the dales and the exhilaration of a view revealed after a short climb.

The new park reclaims as well as reforms; this is a man-made land-scape for human pleasure and activity, characteristics Hargreaves freely admits. Considerations of hydrology, paired with an investigation of the site's history, have generated a sawtooth land pattern that brings

[7-14]
WATERFRONT PARK.
LOUISVILLE,
KENTUCKY. 1990+.
HARGREAVES
ASSOCIATES.
Model.

[7-15]
WATERFRONT PARK.
LOUISVILLE.

[716]
CRISSY FIELD
RESTORATION.
SAN FRANCISCO,
CALIFORNIA. 2000+.
HARGREAVES
ASSOCIATES.

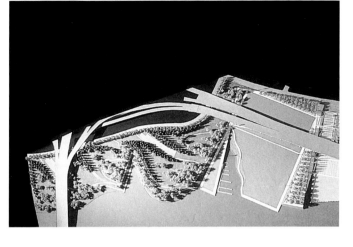

the river deeper into the site. Recalling inlets that existed before the river's regularization, these dentilations also increase the waterfront perimeter and articulate distinct areas of use within the prevalent linear organization. The novelty of the land forms and overall landscape design postpones direct understanding perhaps, instead coaxing the visitor to interact with, and interpret, the park's design as an individual —a lesson of minimalism in sculpture. This is true abstract landscape design, but an abstraction that derives from an understanding of sources in nature while making no attempt to replicate them. The park is a human construct using natural elements where appropriate—and broadcasts just that stance.

The first phase of Louisville Waterfront Park has been completed and the succeeding stages are in progress.[24] Paired with the completion of the Guadalupe River project, it constitutes positive prospects for landscape architecture in the future. The current wetlands restoration project for Crissy Field in San Francisco is more sweeping in its scope and more complex in its mediation of the disparate values of its con- stituents—a set of considerations at least as complex as those con- cerning ecology [7-16]. Some factions of the community sought a complete restoration of the wetlands, with active use by individuals and groups of secondary importance. Others sought to continue the prior uses of the site, which were primarily recreational. And, as one would suspect, the designers felt that a contemporary landscape should reflect current aesthetic ideas as well as social and ecological concerns. The resulting design, at least as it stands today—still in an immature state—reflects quite distinctly these three arenas of consideration.[25] The form of the landscape employs a design strategy of juxtaposition rather than any single entity—perhaps an appropriately complex model for landscapes in the contemporary era.

First, the Hargreaves Associates designs as a group reject the notion of a landscape that emulates nature; they are intended to be "natural, without being naturalistic."[26] They tend toward being green; they are heavily planted; they engage the water in an active way, increasing the length of their edges where shore meets river or bay. But they do not directly strive to recall or replicate natural forms in the manner of the nineteenth-century Olmsted landscape. Although not the words of the designers, one could argue that even nature herself would never produce a "natural-looking" landscape in the park, given the condensed timespan of construction. Construction alters the sweep of process, as a stone tossed into a shallow creek alters its movement. The water continues to flow in accord with gravity and geomorphology, but its nature and its rate of change have themselves altered. Could we not regard landscape design as giving form to natural process constrained by contemporary social and aesthetic conditions, executed in a mere blink in geological time?

These waterfront parks are, without question, designed landscapes of the 1980s and 1990s. While rooted in social use, the varied settings contribute to the whole of the park as a greater project—they are not a series of adjacent playfields or features more significant if taken independently. As designs, these parks evince an art built on history, use, ecology, and, of course, the aesthetics of contemporary form.

At the Parc André Citroën, Gilles Clément installed a *jardin en mouvement* using a neo-Darwinian attitude in which broad-scale seeding was modified over time by the survival of the heartiest species.[27] As it happens, it is the shrubs that have come to predominate, and this one corner of the park today appears solidly planted—evident in these images taken at the time of the park's opening in 1992 [7-17], and in the summer of 2000 [7-18]. For some, perhaps, there is insufficient form apparent in this strategy, particularly as portrayed in photographs. Beyond the camera's frame, however, the frame of the park's overall plan structures and domesticates this wildness and makes it inviting.

Perhaps more surprising are the ecological ideas that propel many landscape designs by Dieter Kienast, who died in 1998. American audiences first encountered Kienast's gardens in a book published by Birkhäuser in 1997.[28] In photographs by Christian Vogt, the Kienast landscape appears black and white, subtly textured and composed, and resting somberly under mostly cloudy skies. In reality, however, one encounters vibrancy, life, and ideas of far greater abundance than those captured on the flattened plane of the printed page.

Without question, Kienast possessed a deft ability for making balanced yet signature compositions, and in some ways his manner conflated the structured spaces of the Italian Renaissance garden with the heavily layered plantings of the English cottage garden—all set in dynamic equilibrium and careful repose.

Ecological understanding underlies many of the Kienast gardens, ideas that remain elusive to the eye. For example, in the restructuring of the

[7-17]
LE JARDIN EN
MOUVEMENT (1992).
PARC ANDRÉ
CITROËN, PARIS,
FRANCE. 1992.
GILLES CLÉMENT

terrace area for the insurance company Swiss Re in Zurich, clearance for the parking level below necessitated a change in terrace level above. Kienast inclined, rather than stepped, the paved surface to collect water runoff, using the gaps between the pavers as drainage channels [7-19]. In areas neither intended for seating, nor draped by the weeping Katsura trees, the gaps were planted with irises almost in the manner of Gertrude Jekyll's terrace garden at Hestercombe.

The horticultural properties of a gigantic collection of plants propelled a garden design for two botanists on the hinterlands around Zurich. In their previous garden, the couple had accumulated nearly 500 species of plants: more or less one of each species. They turned to Dieter Kienast for a new garden that would support aquatic as well as terrestrial species, in a projected number even greater than their then-current collection. The landscape architect described his task as the following: "What does a garden look like to botanists? Moss, loam, solitary bees, handkerchief tree, sand, dragon-flies, rushes, gravel, hedgehogs, cucumber, earth, butterflies. How can these thoughts be formed into a garden?"[29] Alongside the house Kienast aligned in enfilade a series of flat steel tanks for the aquatic plants that led to the rear garden beyond.

Kienast first divided the soils of the rear garden into four distinct strips: gravel, clay, sand, and loam. Species best suited to each of the soils were planted in the corresponding zone. A field of concrete slabs suggesting river ice breaking with the spring thaw overlaid the structured zones of soil producing an antiphonal balance of voices [7-20]. A terrace across the garden accommodated social activity; and as a social ges-

[7-18]
LE JARDIN EN
MOUVEMENT, (2000).
PARC ANDRÉ
CITROËN, PARIS,
FRANCE. 1992.
GILLES CLÉMENT

ture to the community, the garden jumped the rammed earthen wall to offer its pleasures to passersby on the street.

An understanding of horticulture and soils was the basis of the design, and the landscape architect's intervention rested in the superstructure provided by the soils and the fragmented paving. The botanists themselves did the rest. The garden today has somewhere around 650 species—even the owners didn't know for sure. Thus, underlying the jagged patterning that seems so willful is substantial knowledge and structuring. It might be summarized as: Et in arcadia ... eco.

Kienast also experimented with the accumulation of mosses on porous lava stone in a manner that might have shamed the entropic yearnings of Robert Smithson. Within the Swiss Re project a series of misters embedded in a tufa wall dampen its surface and encourage the growth of moss (and one might suspect, mold). Perhaps the pumping system required to maintain the necessary humidity undermines the purity of the idea—for example, would the terrace garner even more respect if the runoff had been used for just this purpose? In fact, Kienast employed just this strategy in a small courtyard for the architectural and engineering firm Ernst Basler+Partner in Zurich. Set almost a story below ground, adjacent to six-story office and apartment buildings, this tiny court receives almost no direct sunlight. Here the tufa forms a retaining wall infiltrated by seepage; over time the moss records the passage of years, its roughness set against the purity and timelessness of the cylindrical water basin fed by piped water run-off [7-21]. Planners and designers who stress ecological processes as the sole

basis of landscape architecture have often disregarded the idea of landscape architecture as aesthetic and cultural practice; those who favor social use have often rejected landscape design as an art. And those who have designed from aesthetic concerns alone have often produced landscapes of stillborn human involvement or neglectful of basic site conditions. In contrast, these projects by Dieter Kienast and Hargreaves Associates propose a potent model for park design, gardens, and more broadly, landscape architecture; one based perhaps more squarely on episodic planning—if one looks to the ideas rather than the particular forms, and time rather than a single moment.

VI.

Social understanding underpins almost all the landscape designs of Georges Descombes. Where to place a bench; how does the figure move? What is the history of the site? How does culture enter the discussion? In this context, Descombes's Parc de Lancy in Geneva, Switzerland, and the commemorative Voie Suisse on Lake Brunnen serve as representative works.

The Parc de Lancy, constructed between 1988 and 1990, lies on the outskirts of Geneva amid housing tracts of relatively high density. The first phase of the design addressed a parcel of land assembled from the sites of three suburban villas from the early part of this century.[30] The terrain slopes steeply from the road toward a shallow ravine; at the lower level vegetation accumulates in greater quantity, producing a strong contrast with the open spaces of the upper zone.

For Descombes, the first step was a careful reading of this rapidly-becoming-urban site. Considerations included the contour of the land and its vegetation, physical surroundings such as the neighboring housing and shops, and of course, patterns of pedestrian and vehicular circulation. To these were added a deeper reading of the park as a place and an institution, attempting to understand not only the super-ficial aspects of the program—rest, relaxation, play, social interaction, contact with the outdoors—but also less obvious ideas about community, behavior, and the history of the site.

The primary strength of Descombes's work is not rooted in its formal appeal—which, one should note, is considerable—but in its integration of history and activity into landscape design and architecture. The invisible, intangible aspects of the design do not fulfill the demands of the camera and yet are deeply felt on site. The limits of the original villa sites, for example, trace the pathways and steps that join the upper and lower portions of the land. Understanding the fatigue that accompanies climbing, and in some cases descent, Descombes positioned benches and seats where they are logically needed—often superimposed upon retaining walls or walls that double as screens against the wind [7-22]. He also investigated, at a level beyond the norm, aspects of children's play. The park's central sandbox, for example, is less a tract of undifferentiated play space than a projection of adult politics onto childhood. In consultation with a child psychologist—paired with his own informal observations—Descombes determined that if the sandbox offers only a single zone, disputes over territory would probably result. To counter this tendency, he divided the play space into several defined zones, each identifiable as distinct [7-23].[31] These psychologically form one unit, however, as does the house in the neighborhood or the neighborhood into the city. The political lessons for the developing child, although unstressed, seem evident [7-24].

The design of the park developed over time as the success of the early phases became obvious and the population density grew. While it is inappropriate to examine all the parts of the project in depth, at least one of the park's principal architectural elements merits further discussion. In a subsequent phase, the park annexed a major parcel of land on the opposite side of the main road, creating problems of linking the land and people. A traffic light was one possibility, although this was impractical; nor did a pedestrian overpass seem to be the appropriate solution. In their place, Descombes proposed a tunnel.[32] Tunnels can be exciting places for children and adults alike, but they

[7-22]
PARC DE LANCY,
GENEVA,
SWITZERLAND. 1988+.
GEORGES DESCOMBES.
Seating in shaded places
beneath a pergola made
from standard green-
house parts.

[7-23]
PARC DE LANCY.
The divided sandbox.

can also be frightening spaces, whose terrifying darkness is compounded by the sudden shift away from the comforting brilliance of daylight into a dismal zone of insecurity. To mitigate potentially negative attitudes, Descombes translated the tunnel into a site of magic, choreographing light levels and modulating the passage from woods to metal tube as a passage from open nature to confined architecture [7-25]. A bridge structure extends the tunnel into the land at either terminus, rendering a negative space positive. In developing the solution, the landscape architect collaborated with the city road department and suggested dividing the traffic lanes above the tunnel, allowing a median between the two directions of traffic. Here a vertical shaft brings light into the heart of the tunnel, just where it is needed most.

As in the big ideas, so in the details. Common materials comprise the basic palette: concrete block left unstuccoed; elements of vernacular greenhouse systems; the metal tubing of drainage culverts. But these everyday elements receive heightened design attention, transforming the common into the special, much as simple bamboo and clay became prized aesthetic objects through the sophisticated transformations associated with the tea ceremony in sixteenth-century Japan.[33] It is not only in his detailing, but also in his sense of detail, that Georges Descombes is such an unusual designer. Like Carlo Scarpa, he understands that a glossy mosaic tile placed in just the right position will reflect light or give color and animate an inanimate surface.[34] In many ways, the design for the Voie Suisse follows in the path of the Parc de Lancy.[35] But in other respects it is a completely independent project that instigated its own way of thinking and its own formal manner. As part of a commemoration of the 800th anniversary of

[7-24]
PARC DE LANCY.
Creative play in the sandbox.

[7-25]
PARC DE LANCY.
The entry of light at mid-tunnel.

the Swiss Confederation, the various cantons proposed a series of memorials and monuments around Lake Brunnen. Quite typically, Descombes eschewed the monument in favor of a less obtrusive presence; instead of a single marker, he proposed a landscape two kilometers in length that would underscore the idea of commemoration by absorbing it into that which could only be Swiss: the Swiss landscape itself. The principal design idea, Descombes once said, was to use a broom.[36] The design of the walk would be less a totally new creation than a revelation of that which had once been, in this case an early nineteenth-century Napoleonic road long derelict and almost invisible.

The strategy would be more about replacement and emplacement than about displacement. Using the "broom," the design team swept away accumulations of vegetation and earth. Where the road needed to be reestablished, small concrete blocks provided support and marked the edge. Where surface drainage threatened erosion, open tracks of stainless steel accommodated the safe passage of water [7-26]. Where the terrain was too steep, or where revised pathways created new intersections, the land was stepped directly and functionally to allow the transition [7-27]. Where railings did not meet contemporary safety standards—in a scenic overlook, for example—new structures overlaid the old [7-28]. The project also restored native plant species where appropriate.

The design team included the artists Richard Long and Carmen Perrin. Long's piece included a letterpress print based on the features of the surrounding landscape; Perrin's contribution was her own particular use of the broom. The site is dotted with erratic boulders, large stones carried by the glaciers far beyond their normal point of deposit. Although most of the local stone is dark gray or black, the erratic boulders are white—their reflective properties kept them relatively cool and underwrote their longer journeys. For Perrin, nothing more was needed than to wash the rocks free of their deposits of moss and dirt. Recast as punctuations and sculptural objects within the landscape, the boulders achieved a heightened presence; but they remained an integral fragment of the landscape nonetheless.

One could discuss the formal brilliance and elegance of all the parts of this design in great detail, but more significant is Descombes's derivation of ideas from the history and form of the site, a poignant model of what the artist Robert Irwin called "site conditioned."[37] By this Irwin implied a sculptural art that could come only from that place at that time under those conditions. There is no way that one can grasp the essence of the Voie Suisse or Lancy landscapes in photographs because it is not concerned with formal structuring. Words may increase our understanding of a landscape, but rarely our experience on site. Because the work extends for a mile and a half, the visitor encounters the landscape sequentially. But this is not a linear landscape as in the ribbon of a road or a single wire. A better reference would be a frayed cable with multiple twisted strands, some of them creating gaps or causing impulses along its length. At certain points the way is physically challenging, causing the visitor to heed the act of walking. In other places, where the slope flattens or a gap in the forest reveals a vista, the event rather than the path controls perception. Underlaying the entirety of this episodic path and movement is the micro-scale of earth, flowers, and shrubs.

Typically, Descombes's regard for behavior, site, history, and structure also informed his design for the 1998 Bijlmemeer Memorial outside Amsterdam [7-29]. Only the more dominant formal elements of the design—the canal, the fountain, and the long concrete retaining walls that double as benches and tables—attract the viewer in images. The sense of longing and absence, however, eludes recording. There is no way to transport a Descombes landscape to another place because the place itself is its most important element. The landscape architect's project here utilizes the eternalized moment of history to inform the making of physical places. The landscape must succeed in the present —social provisions, construction intelligence, aesthetic interest— assimilating the voices of the past with those of the present.

VII.

In this essay, I have tried to establish the possibilities and limitations of landscapes that give primacy to the formal conditions of landscape architecture and patterns that the photograph easily comprehends— and which journals and books then publish for visual consumption. Instead, I would propose that we continue to seek a landscape architecture that engages more fully aspects of the human and natural presence, as well human and natural histories, elevating them through formal dexterity. To provide drainage or seating is only the first response; making that canal or bench beautiful in itself and, perhaps more importantly, an integral contributor—if not outright instigator— for the greater scheme is more crucial. How can any element surpass being

only functional? As Edward Weston once said: "Photograph a thing not for what it is, but for what else it is."[38] So, too, in landscape design.

Of course, all of this must seem like a very simple, very preachy lesson, almost as if this were the first lesson in any landscape architecture curriculum. Perhaps it is; I believe it should be. But I also believe that the lure of the photograph, the manipulated image, and the attraction of the media today have all diminished our interest in these very basic concerns, which is unfortunate. Given the continued evolution of the landscape and its cultural matrix, we should not stop in our attempts to understand their changing content, nor in our search for new manners in which to make them.

Originally published in *Landscape Journal*, Volume 20, Number 2, 2001.

[7-26]
VOIE SUISSE.
LAKE BRUNNEN,
SWITZERLAND. 1990.
GEORGES DESCOMBES.
The path with run-off
channels.

[7-27]
VOIE SUISSE.
Simply executed,
regrading addressed
new changes in contour
or the links between
paths.

[7-28]
VOIE SUISSE.
The scenic overlook.

[7-29]
BIJLMEMEER
MEMORIAL
LANDSCAPE.
AMSTERDAM, THE
NETHERLANDS. 1998.
GEORGES DESCOMBES.
The memorial park is
structured on the foot-
prints of the buildings
destroyed in the airplane
crash. A single fountain
and sheet of water engage
those passing by.

Notes

Trespassing onto such slippery philosophical slopes as those encountered here can only lead to trouble, compounded by the absence of any fixed answers to the questions raised. But as the Zen scholar D. N. Suzuki once said after a particularly animated class discussion: "That's what I like about philosophy: no one wins, no one loses." For intelligently challenging an earlier draft and helping guide my rethinking and revising of the essay I wish to thank the *Landscape Journal*'s three anonymous readers and editor Kenneth Helphand.

1 I have examined this issue in an earlier article in this journal [*Landscape Journal*], Marc Treib, "Must Landscapes Mean?: Approaches to Significance in Recent Landscape Architecture,"

Landscape Journal, Spring 1995, pp. 46–62. [The previous essay in this volume.]

2 A review essay of books concerned with landscape photography centers on this issue: Marc Treib, "Frame, Moment, and Sequence," *Journal of Garden History*, Summer 1995, pp. 126–134.

3 Frederick Law Olmsted and Calvert Vaux, relatively early in the history of photography, used the medium in presenting their "Greensward" design for New York's Central Park. The before-and-after aspect of these sketches probably derives from the "slides" prepared by Humphry Repton for his varied landscapes. See Elizabeth Barlow, *Frederick Law Olmsted's New York*, New York: Praeger, 1972, in particular, pp. 72–73.

4 The English language is unkind here, providing one word to cover

at least two greatly differing applications. *Formal* may address "of or pertaining to form," the adjective drawn from the noun. On the other hand, it may be used as the antonym of "informal," which in landscape terms translates, for example, as gardens planned according to geometry as opposed to those based on nature. In the past and, I fear, also in the current essay, I have found no way to gracefully resolve the issue nor to find other words that convey nearly the same meaning. See Marc Treib, "Formal Problems," *Studies in the History of Gardens and the Designed Landscape*, April–June, 1998, pp. 71–92 [included in this volume].

5 See Clement Greenberg, "Modern Painting," 1960, reprinted in Charles Harrison and Paul Wood, eds., *Art in Theory, 1900–1990*, Oxford: Blackwell, 1992, pp. 754–760.

6 The significance of these distinctions have been dismissed by some critics, however. See John Dixon Hunt, *Greater Perfections*, Philadelphia, PA: University of Pennsylvania Press, 2000, especially pp. 170–175. Here, the author takes issue with the recurring dichotomy in general, and my essay "Formal Problems" cited above, in particular. The world is not words alone, however.

7 One could easily imagine those deprived of vision more heavily favoring the perception of the landscape by other senses such as touch, sound, or fragrance.

8 Dave Hickey, "Bridget Riley for Americans," in *Bridget Riley: Paintings 1982–2000 and Early Works on Paper*, New York: PaceWildenstein, 2000, pp. 8–9.

9 John Brinckerhoff Jackson, "The Word Itself," in *Discovering the Vernacular Landscape*, New Haven, CT: Yale University Press, 1984, note 5.

10 See Marc Treib, "Aspects of Regionality and the Modern(ist) California Garden," in Therese O'Malley and Marc Treib, eds., *Regional Garden Design in the United States*, Washington, D.C.: Dumbarton Oaks, 1995, pp. 5–42.

11 Yu-Fu Tuan, *Topophilia: A Study of Environmental Perception, Attitudes and Values*, Englewood Cliffs, NJ: Prentice-Hall, 1974.

12 The annual awards program of the American Society of Landscape Architects also cites projects in research, planning, and communication. In this paper, I remain directed to answering questions of landscape design that are or could be realized rather than studies about landscapes.

13 "Form in art is as varied as idea itself. It is the visible shape of all man's growth; it is the living pictures of his tribe at its most primitive, and of his civilization at its most sophisticated state. Form is the many faces of the legend—bardic, epic, sculptural, musical, pictorial, architectural; it is the infinite images of religion; it is the expression and the remnant of self. Form is the very shape of content." Ben Shahn, *The Shape of Content*, New York: Vintage Books, 1957, p. 62.

14 See Linda Jewell. ed., *Peter Walker: Experiments in Gesture, Seriality and Flatness*, Cambridge, Mass.: Harvard Graduate School of Design, 1990; Peter Walker, "The Practice of Landscape Architecture in the Postwar United States," in Marc Treib, ed., *Modern Landscape Architecture: A Critical Review*, Cambridge, Mass.: MIT Press, 1993, pp. 250–259; and Marc Treib, "Motifs, trames et structures" (The Place of Pattern), *Pages paysages*, no. 4, 1992–1993, pp. 128–133.

15 The extreme example of this phenomenon is the work of Andy Goldsworthy, whose sculpture is often more powerful in photographs than in actuality—and at

at times seems to demand to be photographed, given the ephemerality of the work. The play between the installation and the defining rectangle of the photographic frame heightens the presence of the work and removes it from its greater context —of which, often, it is only a very small part. Rather than reading a spiral of colored leaves against the irregular patterns of natural elements, for example, we read it against the photographic frame infilled with the irregular patterning of nature. See Andy Goldsworthy, *Hand to Earth: Andy Goldsworthy Sculpture 1976–90*, New York: Abrams, 1990.

Richard Long's documentation of his walks differ in that the photograph is a mnemonic device for the artist and a narrative device for the viewer. The photograph recalls the event as marked by Long's construction rather than as acting as part of the work itself. See R. H. Fuchs, *Richard Long*, New York: Guggenheim Museum, 1986; and Richard R. Brittell and Dana Friis-Hansen, *Richard Long: Circles, Cycles, Mud, Stones*, Houston, TX: Contemporary Arts Museum, 1996.

16 For a discussion of the ideas behind the park and its design process, see Alistair T. McIntosh, "Burnett Park," in Jewell, *Peter Walker*, pp. 30–37.

17 This project is presented in Dan Kiley and Jane Amidon, *Dan Kiley: Complete Works*, Boston: Bulfinch, 1999, pp. 106–112; and *Dan Kiley: In Step with Nature; Landscape Design II, Process Architecture* 108, 1993, pp. 46–54.

18 Tom Maver, managing landscape architect, in conversation with the author, Solana, November 1996.

19 While this essay was in press, Kenneth Helphand informed me that the proscription of tree planting no longer directs current practice. I thank him for bringing this to my attention. At the time

that Byxbee was designed, however, these restrictions were active.

20 I would imagine that few clients would be thrilled by the idea that their garden will erode over time, or that distinctions of bed and lawn might disappear with the spread of weeds or native grasses. The distribution of grasses and wildflowers through reseeding, or the growth of volunteer trees over time, might constitute beneficial examples of applied entropic order, however. The most important of Smithson's essays to focus on this subject is "A Tour of the Monuments of Passaic, New Jersey," (originally published in *Artforum*, December, 1967), in Nancy Holt, ed., *The Writings of Robert Smithson*, New York: New York University Press, 1979, pp. 52–57. As early as 1971, art historian and psychologist Rudolph Arnheim had published *Entropy and Art: An Essay on Order and Disorder*, Berkeley, CA: University of California Press, examining this natural drift in relation to art production.

21 These forms, in fact, have become a signature element in the Hargreaves Associates landscape, appearing in all the waterfront designs, but are less apparent here—at least so far—due to thick natural vegetation. Plantations of palms, poplars, and conifers reinforce the patterns set at ground level. While some areas have already been completed, work will continue for half a decade.

22 The site was first slated for redevelopment in the mid-1980s.

23 There will be some few structures for dining or recreation, but the hotels and housing of the original proposal will have to address the river from beyond the limits of the park.

24 *A second phase, a children's park area, opened in summer 2003.*

25 As of January 2001.

26 George Hargreaves, in conversation, Cambridge, Massachusetts, October 1997.

27 Gilles Clément, *Le Jardin en mouvement: de la vallée au Parc André Citroën*, Paris: Sens & Tonka, 1994.

28 Dieter Kienast, *Gärten Gardens*, Basel: Birkhäuser Verlag, 1997.

29 Ibid., p. 168.

30 For an overview of this project, see Giordano Tironi, ed., *Il Territorio Transitivo/Shifting Sites*, Rome: Gangemi editore, 1998.

31 What should happen if there are more than four play groups, however, is open to question.

32 *For an excellent discussion of Descombes's ideas at the Parc de Lancy, see Sébastien Marot,* Urbanism and the Art of Memory, *London: Architectural Association, 2003, pp. 58–78.*

33 This elevation of common objects to the status of high art depended on erudition in taste and painstaking selection or reworking. As the museum today recontextualizes ethnographic or artistic production, the teahouse removed the everyday object from its mundane context, elevating its aesthetic status by an appreciation of its simple values. In actual practice, the tea masters more commonly developed their own arts with a nod in the direction of the everyday rather than extensively using truly common wares.

34 In his use of the judiciously placed reflective metallic tiles, Descombes recalls elements of the 1969–1978 Brion Cemetery, San Vito d'Altivole, and the 1973 garden for Palazzo Querini-Stampaglia in Venice, both designed by Carlo Scarpa.

35 A book documents the process and the elements of the project: *Voie Suisse, l'itinéraire genevois: De Morschach à Brunnen*, Fribourg: Canton de Genève, 1991.
Descombes considers the book a part of the project, the landscape of which was understood to be ephemeral.

36 Conversation with author, Geneva, July 1999.

37 Robert Irwin, *Being and Circumstance: Notes toward a Conditional Art*, Larkspur Landing, CA: Lapis Press, 1985, pp. 26–27.

38 I have not been able to locate the exact source of this quote although I suspect it comes from Nancy Newhall, ed., *The Daybooks of Edward Weston*, New York: Horizon Press, 1966.

Evocative Parallels:
Japan and Postwar American Landscape Design

2002

I.

Many period images of the postwar California garden display two
distinct characteristics: a vocabulary derived from the modern arts
of painting and sculpture, and a sensibility that appears to be at least
mildly Asian [8-1]. Of the former, the biomorphically curving outlines
of swimming pools and planting beds, checkerboarded wooden decks,
the zigzag profiles of screen walls or perimeter planting, all smack of
the fine and applied arts dating from the 1920s onward. But if these
formal vocabularies wear the garb of the artistically new, the sense
of reduction and defined focus—if not the use of exposed beam ends
and plant species such as bamboo and Japanese maple—suggest the
influence of Japan and the Far East. But do these hints of aesthetic
relations really demonstrate influence, or do they just illustrate
provocative parallels in aesthetic imagination?

To better inquire into the nature of this question, we might first try
to define just what constitutes "influence." Running to the dictionary
is the sanctuary of every academic, as if the dictionary could better
tell us what one hopes we already know. Instead, let me suggest that
an influence, at least in terms of the design professions, is an idea—
theoretical and abstract, or manifest and perceptible—which directs
in some way the development of a design. Of the two types, physical
models—transferred or transformed—are more easily identified,
especially in architecture. Le Corbusier's forms reincarnated into the
work of Richard Meier offer an obvious example. But we need also
include theoretical ideas that lie behind the forms because influence
can trace to many fields without showing any literal manifestation.
As a prime example here we might cite the admitted influence of the
transformational linguistics of Noam Chomsky on the architecture of
Peter Eisenman.

Loans may be unconscious as well as conscious. Meier's borrowings
from Le Corbusier are aware, overt, and there for the world to see.
But all of us—as writers and/or designers—have absorbed some
idea or form without knowing, only to have it return at some later
date as seemingly fresh and original. These can ultimately form
extended chains of influence, at times intersecting, at other times
parallel. Constantin Brancusi, for example, created the *Table of
Silence*, a 1937 memorial park at Tîrgu Jiu in Romania [8-2]. The
round table with its goblet seats contrasted roughness and solidity
with the purity of its circular geometry: a beautiful sculpture that may

also be used to support the body and human activities. The similarities to Pierre-Émile Legrain's stone table at the Tachard garden in La Celle Saint-Cloud, France—with its "hypnotically off center" tree trunk —are so close that it would seem that Brancusi must be its source [8-3].[1] In fact, the garden dates to circa 1924, almost fifteen years before the Romanian sculptor returned to his homeland to execute his own work. Influence is difficult to ascertain here; perhaps parallel, spontaneous creation was more likely.

From that point on, however, I believe we can be more certain in the transfer and transmutation of the Brancusi sculptural idea. One version informed Isamu Noguchi's seating at Unesco House in Paris from 1958 [8-4]. This is not a copy of a form, but the adoption of an *idea*: seat/sculpture. We know that the American sculptor Scott Burton well knew Brancusi's work—he even curated a show of the Brancusi sculpture bases at the Museum of Modern Art in New York; like the sculptures, Brancusi always created their supports himself.[2] Burton's 1986 seating for the Equitable Life Insurance Company in New York appears as the product of a cross-fertilization by Brancusi and the Milan furniture group Memphis. And finally, the table and seats for the 1984 TEPIA building in Tokyo by Fumihiko Maki appear more a direct homage to Brancusi than a surreptitious borrowing or influence [8-5].[3]

In *The Anxiety of Influence*, literary critic Harold Bloom outlined six stages that inform the development of the poet, a development tracked through the assimilation of influence.[4] Over time, the poet dependent on the influence of those who had come before him or her is transformed into one marked by independent thinking. A dialectic operates here to some degree, but Bloom's recounting also suggests that no one is truly original, that no one starts from a tabula rasa, that no one is free from influence. Bloom's discourse is so dependent on the written form, however, that it appears to be of only limited application to design with its manifold dimensions and sources, both abstract and manifest.

II.

With this (admittedly somewhat shaky) groundwork laid, let us return more specifically to Western landscape architecture and the putative influence of Japan. For the most part, specific references are brief and relatively few. One of the earliest and least contestable links is found in the early writings of Christopher Tunnard. Born in Canada,

[8-2]
TABLE OF SILENCE.
TÎRGU JIU, ROMANIA.
1937. CONSTANTIN
BRANCUSI.

[8-3]
TACHARD GARDEN.
LA CELLE SAINT-CLOUD,
FRANCE. CIRCA 1924.
PIERRE-ÉMILE
LEGRAIN.
LA SALLE FRAÎCHE.

[8-4]
UNESCO HOUSE.
PARIS, FRANCE,
SEATING, DELEGATES'
TERRACE. 1958.
ISAMU NOGUCHI.

[8-5]
TABLE AND CHAIR,
TEPIA. TOKYO, JAPAN,
TABLE AND SEATS. 1984.
FUMIHIKO MAKI.

but schooled in horticulture in England, Tunnard was one of the first to formulate a new landscape architecture, one in accord with the contemporary era and its conditions. How Tunnard made the leap to Japan is still not established conclusively, but one *can* trace a path of influence in his writings. In the preface to his 1938 *Gardens in the Modern Landscape* Tunnard credits many of his modernist leanings to Frederick Etchells and his understanding of things and aesthetics Japanese to Bernard Leach.[5] Etchells was a painter and printmaker and a member of the influential British Vorticist group, which also included at times Edward Wadsworth and the artist/critic Wyndham Lewis. Not incidentally, Etchells translated in 1929 Le Corbusier's key work *Towards a New Architecture*, a book that in many ways served Tunnard as a model for his own polemic. Bernard Leach, on the other hand, was a potter who first traveled to Japan in the 1920s, and under the tutelage and friendship of potter Shôji Hamada engaged in the fertile crossing of European and Japanese ceramic techniques across a number of decades.

Perhaps it was through Leach that Tunnard learned about the innovative landscape designs of Sutemi Horiguchi. One can hypothesize that the English potter encountered the Japanese architect through the *mingei* (folk art) circle in Tokyo and that it was he who brought the small monographs Horiguchi published on each of his completed works—with dual texts in Japanese and German—back to England and to Tunnard. No matter the origins or the specific vehicles, by 1934 Tunnard had published his first article on Japan in *Landscape and Garden*, remarking on the influence of Japan upon English landscape design.[6] In some ways the article is a bit naïve—for example, assigning the essential green-ness of the English landscape garden to roots in Japan, where floral color is carefully restricted. "Now, while the trend in all art is slowly moving towards an acceptance of form, line and economy of material," Tunnard wrote, "some of our gardens are beginning to show signs of this change of opinion. There can be no doubt of the influence of Japan here." Or can there? Tunnard discusses the role of stones in the Japanese garden, particularly their use as its "skeleton": "a sure way of obtaining balance and proper relationship of the separate parts." He says almost nothing specifically about the Horiguchi project with which he illustrates his article, however; doubly interesting is that no historical Japanese gardens appear to bolster his assertions.

By 1937–1938, when Tunnard first published *Gardens in the Modern Landscape* in serial form in *Architectural Review*, the Japanese garden had become his preferred model for a new Western garden genre. Tunnard equally dismisses the formal manner of the French classical garden and the decisive informality of the English landscape garden. For contemporary life, asserts Tunnard, there should be a new manner —the *empathic* garden. This garden "technique," as Tunnard termed it, was characterized thus: "The banishment of the antagonistic, masterful attitude toward Nature, of excessive symmetry, a recognition of the value of tactile qualities in plant material, grasp of rhythm and accent, contribute to the supple and fluid adaptation of the site, which is the landscape architect's chief arbiter of design."[7] Key to this was a unity of garden and habitation. In the Japanese garden, "all is restrained, calculated and under control. No great waves of vegetation beat upon the building; man shows his respect for Nature by allowing her admission to his scheme of planning, but sees to it that the respect is mutual."[8] He admonishes his reader: "Let us absorb the Oriental aesthetic. By doing so we shall help to rid ourselves of individualism in art and gain identity with art and life."[9]

Tunnard's writing exerted an enormous influence on the young Garrett Eckbo, pursuing graduate studies at Harvard University about the same time as Tunnard's polemic appeared in print. Likewise, Lawrence Halprin, at Harvard some years later, also fell under the Canadian-Englishman's spell. In 1986, Halprin revealed that he had read Tunnard's manifesto in 1940: he "decides on the spot to study design in architecture, with an emphasis on landscape design."[10]

Both Eckbo and Halprin were at least half a generation younger than the landscape architect Thomas Church, who had already assumed a position of prominence in the San Francisco Bay Area design community. There is no direct connection between Church and Asia, and yet in certain places shared forms, plant species, and sensibilities are evident (also, we must not discount the possible contributions from members of his staff). Church's early work owed more to the formal tradition than to either the modern arts or Japan. However, in the rear sections of some early gardens, the formal design dissipated into naturalness —a lesson from Japan, or Frederick Law Olmsted, or from Humphry Repton beyond him?

Works from the 1940s, however, employ features associated more directly with Japan: for example, the boulder used as an ornamental element, or as a punctuation to the garden field. This regard for stone is endemic to China as well as Japan, of course, where the presence of the *kami*, or deities, could be signaled by unusual topographic formations, indeed, even one notable rock. The celebrated Donnell garden of 1948 showcased a variety of these boulders which, as in many Japanese gardens, structured the landscape. But despite the visual strength of the rocks, it was the swimming pool and the paving that prevailed, and not the boulders left from the site preparations. During the course of design Lawrence Halprin, who was project designer for the garden, suggested that they use a large rock in the pool as a centerpiece for the composition. Church, instead, opted for a tall biomorphic sculpture by Adaline Kent, most of which remained below the surface of the water. As a result, the garden was to smack of modernity in Western style rather than any manner associated with the East.

In certain gardens, such as the 1960s Mudd garden in Hillsborough, rocks run rampant [8-6]. One reads the pool area as a terraced slope into which the swimming pool has been inserted—with boulders strewn along, across, and around the pool as if by erratic geological movement. One is tempted to assert that this gesture betrayed a Japanese influence, and in fact it just might. But if one compares the rock work and placement in the traditional gardens of Japan—the use of clustering, spatial balance, and definition of edge—there is precious little that recalls directly the historical landscapes of Kyoto. Should we not better term these "Modern-Asian-American," if any categorization is required? It would seem that at this stage in the process of borrowing and assimilation, a novel hybrid—novel in its own way—had already evolved.

And if there is an Asian breath behind the design, might Church's (or Halprin's) sculptural use of rocks have entered Church's work via Frank Lloyd Wright rather than directly from Japan? At Taliesin West, built in 1938 outside Phoenix, rocks taken from the site were formed in concrete to make the walls; singular boulders serve sculpturally and to instigate contemplation [8-7]. Perhaps Halprin, or Church, knew more about Wright than about Japan; perhaps they did not consciously recall that Wright may have received his inspiration from Japan decades earlier when the strongest influences in Wright's work were apparent.

Normally Wright was reticent to admit he had had any influences whatsoever, so the position he outlines in *The Future of Architecture*

comes somewhat as a surprise: "Artists, even great ones, are singularly loath to quote sources of inspiration—among lesser artists ingratitude amounts to phobia. No sooner does the lesser artist receive a lesson or perceive an idea or even receive the objects of art from another source, than he soon becomes anxious to forget the suggestions, to conceal the facts, or, if impossible to do this, to minimize, by detraction, the 'gift.'"[11] He mentions no particular source, of course, demonstrating once again that it is dangerous to trace influence with any degree of confidence.

Eckbo's Harvard classmate James Rose provides us with a set of relationships and transferences more easily established. In certain writings Rose denies the possibility of working truthfully beyond one's

native cultural sphere. In *Gardens Make Me Laugh*, for example, he tells of the potential client who telephones inquiring whether or not Rose designs Japanese gardens. "Sure," he replies jovially, "where in Japan do you live?"[12] Yet despite this denial, the influence of Japan pervades much of his work—at least that is how I read it, and he inferred it. His denial is not as assertive in later writings, and the designs more transparently refer to Japanese architecture and their accompanying landscapes. His use of horizontality, the shoji, the veranda, and the direct connection between inside and out all even *look* Japanese—despite his continued reciting that he was interested in the Japanese spirit and not its forms. Rose cites T.D. Suzuki on Zen but nothing on garden making. How should we interpret these facts?

In later years, Rose spent extended periods of time in Japan, by his own telling at least two trips a year, and a full year's stay in 1973.[13] And in his last book, *The Heavenly Environment*, the subject of Japan and Japanese thinking recurs in many parts of the texts, starting with the opening pages. His later work reads as openly as a written text on influence. For example, the transformation of Rose's own compound in Ridgewood, New Jersey, reflects—like a self-portrait—the increasing influence of the East [8-8]. He might have denied it; but what greets the eyes denies the denial.

Garrett Eckbo and Japan also maintained a cordial relationship, which included the American's living in Japan as visiting professor at the Osaka Prefectural University in the early 1970s. Eckbo maintained a true appreciation for the accomplishment and the beauty of the Japanese garden, but there is little in his work that can be assigned to Asian influence, at least not at the superficial level. Unlike Church, the boulder was never a major element of Eckbo's designs, although in some fountains the giant slab of stone did figure prominently (perhaps more the influence of Lawrence Halprin than Japan, however). In his gardens, he looked more resolutely West than East, especially in his park and institutional work. Modern art and sculpture gripped him more tightly than any Asian garden forms, and one rarely finds in his work the void, that quiet center so characteristic of great Kyoto gardens such as Shôden-ji or the garden of the Silver Pavilion, Ginkaku-ji.

If Japan exerted a spell at all, it was to evoke an appreciation of what made a landscape design great. He wrote in 1950: "What is relevant to us [about Japan]"—and what was relevant to Eckbo in his own work—

> is what we see in the physical result: the restraint, the calculation, the control; the orderly sculptural use, in infinite variety of arrangement, of rocks at every scale from sand to boulders;...the endless delicate training of trees and shrubs to produce an exact line or weight; the constant ingenuity in the use of stray materials, straw, sticks, stones, sand piles; the careful modeling of ground, shrub, and rock forms into continuous sculptural wholes.[14]

In *The Art of Home Landscaping* of 1956, Eckbo again cited Japan, but here he stressed that the understanding of mass and void was the ultimate lesson to be learned—not the use of cultural symbols like the maple or the lantern. In sum, Eckbo extracted and abstracted from his source. He may have had an "influence" from Japan, but there is virtually nothing that *looks* Japanese or relies in any way upon Asian iconography.

Curiously, Christopher Tunnard's first—and to my knowledge only —travels in Japan were as late as 1960, when he attended the World Design Conference there. A text on the garden of the Silver Pavilion developed from that visit.[15] In the text, Tunnard discusses the approach and movement through the garden, the formality of the sand mound and cone, the concealed boundaries of the garden, and even the symbolic associations with species such as the plum, the cherry, and the pine. But no matter their charm and seduction, Tunnard implies that Japanese gardens should stay in their country of birth: "Japanese gardens are not for export. They are complicated and require the most minute attention to detail for their upkeep. Further, our backgrounds are completely unsuitable as a foil." He concludes: "But, gardening the world over would undoubtedly improve if the Japanese art of relating and varying natural material were properly studied. This is their great contribution."[16] Twenty-six years had passed since Tunnard had published his first thoughts on the Japanese garden design, and yet he came remarkably close to what he had written in 1934: "It seems clear that copyism, even if desirable, cannot be carried out by those of another race."[17]

III.

Given these somewhat random citations informing the work of a handful of the leading twentieth-century landscape modernists, can we distill any conclusions about influence, its definition and its effects? First, I would point to the danger in trying to assign cause and effect to specific works, or even more broadly, to the thinking of a designer. As Harold Bloom pointed out, the stage in a poet's career during which the influence arrives is often more critical than any inherent predilection toward influence. When a strong designer matures, he or she normally comes to absorb external influences old and new, ultimately to speak with an independent voice. That original influence, if we must use the word, appears to have all but disappeared given its reduction in status to but one of many. Le Corbusier once said that he was like a lion, constantly devouring all sorts of sources. Asking what he had eaten made little sense; for he had eaten from too many sources.

And what of the case of Japan? That too remains problematic. Certainly, Japan has exerted a huge influence on Western arts and design, from at least the mid-nineteenth century on. A major study like Clay Lancaster's *The Japanese Influence in America* of 1963 clearly establishes the impact of Japanese thought on our aesthetics, architecture, painting, applied arts, and gardens [8-9, 8-10].[18] Lancaster's study also under-scores the fact that influence need not come directly from a similar medium, like garden making. It may arrive from the fine or applied arts, in this particular case from Japan. Or even farther afield, it may appear through chains such as these: Japanese prints influence Western painting, which develops into abstraction, which influences postwar garden design. Does that constitute a Japanese influence or do we need to have a readily demonstrable sign of a direct connection?

Thus, it would seem that structures drawn from literary criticism do not easily apply directly to design. Instead, I would propose four manners by which influence is received and used, each with less conspicuousness. These are: *replication*; *citation*; *adaptation*, and *abstraction*. *Replication* implies a direct copy, like the replica of the Ryôan-ji garden at the Brooklyn Botanical Garden.[19] This constitutes a nearly one-to-one correspondence between the original and the re-enactment. The use of boulders sculpturally in the Thomas Church projects cited above represents *citation*: using elements of the original source in part, rather than as a whole. *Adaptation* may borrow theoretical or practical ideas, but transform them into new materials. The rocks of the previous stage may be translated into sculpture, or a void once sand may now become water. Influence in the last stage, *abstraction*, is the most difficult to ascertain (unless by the personal admission of the designer), since the sources may in fact be multiple. Garrett Eckbo's handling of rock and plant may illustrate this category. But at this point some may see influence where others do not. Could Eckbo's sense of space, of playing mass against void, of use of asymmetrical balance, be traced to Japan? Or to modern art? Or to his own development through years of active practice?

Perhaps, perhaps not. Other than in the replication phase, influence is notoriously difficult to establish with certainty. I feel more comfortable proposing Japan as more a mirror than a lens. James Rose, too, used this metaphor when he wrote: "But the meeting [of East and West] is like two cultures viewing each other without realizing there is a mirror of self-image between them, and until that mirror is removed or broken, one sees self instead of whatever is on the other side of the looking glass."[20] The American architect Ralph Adams Cram, with a more than subtle Gothic predilection, traveled to Japan around the turn of the century. There he found architecture with a clear expression of structure, which he read as substantiating the Gothic way of building. That the Japanese architectural style was so different, or that spaces tended to be horizontal rather than vertical, seems to have mattered little to him. Walter Gropius arrived quite late, after the war, but he found at the Katsura Villa virtually everything that the Bauhaus *Schule* had sought: modular planning, free spaces, flexibility, systematic building construction. Here we see the crux of the problem: same source, radically differing interpretation—and potential applications.

These two architectural cases suggest that influence is only what you make of it; and that in the borrowing process, the influence can become so distorted or so weakened that it serves merely as a stimulus for creative transformation rather than as a subject for replication. As we must admit the importance of transformation *after* influence (in all but blatant plagiarism), so we might also notice that in most cases creative distortion, rather than literal citation, instigates the more innovative form or productive use. At that point we might very well question, one final time, the importance—or even the very existence—of influence as a significant factor in landscape designs, whether it comes from the east or from the west, from the past or from today, from other gardens or other fields.

Originally published in the *Consultants in Landscape Architecture Journal* (in Japanese), Number 152, January 2002.
An earlier version of this paper was presented at the 2001 Annual Meeting of the Society of Architectural Historians.

[8-9]
KINKAKU (THE
GOLDEN PAVILION).
KYOTO, JAPAN.
CIRCA 1394,
REBUILT 1950s.

[8-10]
DOULLUT HOUSES.
NEW ORLEANS,
LOUISIANA. 1905.
The Golden Pavilion
meets the steamboat.

Notes

1 For an overview of the Tachard garden design, see Dorothée Imbert, *The Modernist Garden in France*, New Haven, CT: Yale University Press, 1993, pp. 113–123.

2 The exhibition, "Artist's Choice: Burton on Brancusi," was shown at the Museum of Modern Art in New York, 7 April–28 June, 1989.

3 *Illustrations of these designs, including the Scott Burton work, appear in Marc Treib,* Noguchi in Paris: The Unesco Garden, *San Francisco: William Stout Publishers, 2003, pp. 85–88.*

4 Herbert Bloom, *The Anxiety of Influence*, New York: Oxford University Press, 1997.

5 Christopher Tunnard, *Gardens in the Modern Landscape*, London: Architectural Press, 1938.

6 Arthur C. Tunnard, "The Influence of Japan on the English Garden," *Landscape and Garden*, Summer 1935, pp. 49–53. The principal illustrations are Sutemi Horiguchi's Kikkawa house and garden, most of which would later appear in *Gardens in the Modern Landscape.*

7 Tunnard, *Gardens in the Modern Landscape*, p. 105.

8 Ibid., p. 90.

9 Ibid., p. 92.

10 Lynn Creighton Neal, ed., *Lawrence Halprin: Changing Places*, San Francisco: San Francisco Museum of Modern Art, 1986, p. 115.

11 Frank Lloyd Wright, *The Future of Architecture*, New York: Horizon Press, 1953, p. 110.

12 James Rose, *Gardens Make Me Laugh*, Norwalk, CT: Silvermine Publishers, 1965, p. 83.

13 James Rose, *The Heavenly Environment*, Hong Kong: New City Cultural Services, 1987, p. 7.

14 Garrett Eckbo, *Landscape for Living*, New York: Duell, Sloan, Pearce, 1950, p. 16.

15 The typescript, now in the Sterling Library at Yale University, New Haven, Connecticut, does not tell if or where it was published, however.

16 Christopher Tunnard, "The Garden of the Silver Pavilion," p. 3.

17 Tunnard, "The Influence of Japan," p. 50.

18 Clay Lancaster, *The Japanese Influence in America*, New York: Walton Rawls, 1963.

19 I never visited the garden, which I am told no longer exists, but it was rumored to have been built as the reverse of its Kyoto prototype.

20 Rose, *The Heavenly Environment*, p. 8.

The Measure of Wisdom: John Brinckerhoff Jackson

1996

In his last years, John Brinckerhoff Jackson spent the morning hours working as a gardener-custodian. He customarily wore jeans and a T-shirt in summer, or in winter, a flannel shirt, and leather pants and jacket [9-1]. He sought the company and commentary of those he encountered while working or shopping rather than those academics who held him in such high esteem. Visually, Jackson hardly evoked the air of a senior, well-respected scholar, and many of his academic friends were puzzled by his quotidian blue-collar labors. In all probability, his efforts were driven by the need for interacting daily with those who comprise the keel of American society, and the American cultural landscape of which Jackson was such a keen observer.

"Brinck" Jackson—as he was known in the academic sector and to older friends—engaged the full range of American life. It would be tempting to cast him as a patrician among hoi polloi or a simple person among academics or the social elite. But like his work, he was not so easily classified. Put simply, Jackson believed that understanding the people of the United States and their built environment (or cultural landscape, as he preferred to term it) comes not by examining any one segment in isolation, but only through the broadest embrace. His life was dedicated to acquiring this understanding.

John Brinckerhoff Jackson was born in Dinard, France, in 1909 and spent his early years in Europe. Schooled in Switzerland, he returned to the United States when his parents separated. Even as a child, his powers of observation were unusually keen (and so they remained, both literally and figuratively—he read without glasses until his passing). Jackson attended Deerfield Academy, Choate, the University of Wisconsin, and later Harvard, graduating there with a B.A. in history and literature in 1932. While he rarely referred to his eastern education except in mock derision, he did recall the lasting influence of Irving Babbitt on forming his aesthetic tastes. For Jackson, art was to be edifying, something that elevated feeling and belief rather than constituting a commentary on reality. More unusual, however, was his essential rejection of the idea of the landscape as form ripe for aesthetic judgment; instead it was principally the setting for living that reflected the relationship between people and the land. Hoping to become a cowboy, after graduation from college in the late 1930s he traveled west for the first time, to his uncle's ranch in Cline's Corner, New Mexico. The American entry into the Second World War provided him with a new mission, however, and instigated his departure from the plains and mesas of the Southwest. The American terrain had prompted, perhaps, his examination of geography; the war honed his abilities to read it.

In "Landscape as Seen by the Military," Jackson tells of the urgent need to understand the battlefield terrain and the reasons for its significance. During training, he had assumed that "the occupants of the environment were animated by one very clear purpose: to hold onto it as long as they could. We learned to study the environment only insofar as it might help or hinder the carrying out of that purpose."[1] After exposure to battle in Normandy and North Africa, and prompted by the military's essential disinterest in the particularities of the landscape beyond those detrimental to troop movement and acquisition, Jackson reached another conclusion: "Finally, the landscape ceased to be an empty impersonal stage and became part of a whole way of life, a place where men and environment were in harmony with one another and where an overall design was manifest in every detail."[2] Every landscape was an essay. No landscape could ever escape its time, or, most certainly, reflecting its people. He concluded:

> Every landscape of any size or age has a style of its own, a period style such as we discern or try to discern in music or architecture or painting, and a landscape true to its style, containing enough of its diagnostic traits, whether it is in Appalachia or Southern California, can give an almost esthetic satisfaction.[3]

As he had acquired fluency in French and German, Jackson had learned to sketch at an early age—a practice he would continue throughout his life—and drawings of the land, settlement, towns, and farms recorded these patterns of human intention [9-2]. He paired on-site documentation with forays into local libraries, sorting through provincial histories and diaries—and any other documents that narrated the story of the landscape.

After the war, Jackson returned to New Mexico and sought a vocation. A fall from a horse, a broken leg, and a period of enforced convalescence ended his career as a wrangler. He turned, instead, to the study of landscape that his martial experience had first encouraged. In 1951 he founded, and began to edit *Landscape*, "a journal of human geography," conceptually based to a large degree on the notions of human geography advanced by Marc Bloch and the French Annales School. Recent research by Helen Horowitz and Paul Groth has ascertained that Jackson wrote the early issues entirely by himself, concealing his authorship by using a host of pseudonyms.[4] Gradually, the focus on

the Southwest dissipated—perhaps due to the lack of qualified contributors besides the editor himself—and the purview of the journal
broadened, attracting the attention of geographers, sociologists,
architects and landscape architects, and historians. Although standing
resolutely outside the university arena, he also began to acquire
academic standing.

Jackson never saw the place without its people, and in *Landscape* he
provided the forum for architects and architectural historians as well
as botanists, geographers, and anthropologists such as Edward T. Hall.
Jackson believed that the High Style monument was necessary to
complete a comprehensive regard of the environment, but he rarely
singled out these buildings in his own writings. While their stories and
its clients often came from an elite group at odds with the vernacular,
they were nonetheless significant components of the landscape.
Contributors to *Landscape* included László Moholy-Nagy, Sigfried
Gideon, Lewis Mumford, Bruno Zevi, and Reyner Banham. During
the 1960s, Denise Scott Brown, Charles Moore, Donlyn Lyndon,
and others used the magazine to discuss topics as wide ranging as
Levittown and The Strip, and the now near-mythical "sense of place."
While today the publication of essays on these topics might not surprise us, they found little following in professional design journals at
the time.

There is little need to rehearse the hundreds of articles that issued
from J. B. Jackson's pen (he wrote by long hand all his life) during his
seventeen years of editing *Landscape*, or his lectures, or his countless
other articles. It is informative to cite the range of these writings,
however. Jackson was, for example, the first critic to look seriously

at the architecture of The Strip, at a time when the American roadside environment was moving toward respectability, and the very things that caught Jackson's eye were disappearing. In "Other Directed Houses" (first published in 1956) he wrote:

> [W]e must give these roadside establishments their due. They are entitled to their day in court, and so far they have not had it. Many have experienced driving for hour after hour across an emptiness — desert or prairie — which was *not* blemished by highway stands. How relieved and delighted one always is to finally see somewhere in the distance the jumble of billboards and gas pumps and jerry-built houses. Tourist traps or not, these are very welcome sights... The gaudier the layout the nicer its seems, and its impact on the surrounding landscape bothers not at all.[5]

Of course, the scale of the construction and the scale of the natural terrain established the curious aesthetic symbiosis of shanty and landscape. But the sanitized landscape, in Jackson's eyes, would be far worse, reflecting far fewer of those values Americans hold dear, individualism primary among them.

He shunned the academic paraphernalia of sources and footnotes, approaching his subject as a true essayist, and his prose is a model of clarity and economy, avoiding the fashionable sources—though he had probably read them all—opting instead for a written form that best expressed the material he wished to discuss. Perhaps the most outstanding illustration of this empathetic approach was "The Westward-moving House" of 1953. Here Jackson rejected the rigors of conventional exposition and told his tale as a short story, fictionalizing the lives of three families as the point of departure for explaining the landscape that both embraced and was formed by them. These people have names, they have the accoutrements necessary to living, they have activities, and, of course, they have a house. We learn first of early New England settlers, the family of Nehemiah and Submit Tinkham; we jump to the western expansion, to Pliny Tinkham in Illinois; we end in the Sunbelt, in Bonniview, Texas, in a tract house occupied by Ray Tinkham and his family. Ray had been a cattle rancher, we learn, but the discovery of abundant underground water allowed him to shift over to planting the land to "sorghum or castor beans, depending on the market."[6] One might dismiss Jackson's form as facile cleverness, and oh my God! Where are the footnotes? But by integrating his factual material with a narrative, by creating a fictive landscape—in Jackson's view the interaction of people and place—he forged a literary form

[9-3]
THE JACKSON HOME
AND LANDSCAPE.
LA CIENEGA, 1986.

that better expressed his views than the scholarly essay. To his credit, one recalls the minutiae *because* they are a meaningful part of a coherent view; indeed, as the real landscape is itself.

In "The Necessity for Ruins," Jackson examined more closely the landscape as a repository of memory, and it remains one of his most prescient and enduring essays. Unlike the architectural historian, who tends to focus on the discrete building or commemorative ensemble, Jackson tackles the question of the commemorative landscape as a whole. He suggests that today "history means less the record of significant events and people than the preservation of reminders of a bygone domestic experience and its environment."[7] He is puzzled that the current "novel interpretation of history carries over from the museum or private collection into the wider rural and urban landscape." A comparison between memorials in the United States and in South America outlines—formed as another narrative—how each recasts its environment as a setting for commemoration.

> The town of Centerville suddenly realizes that it is approximately one hundred fifty years old. (No one knows the year when the first settler arrived, or who he was.) A mass meeting is held in the high school gym to discuss how to celebrate the event. It is finally agreed that it would be nice to feature the arrival of the first train, sometime in the 1850s, and the Indian raid of 1847, and at the same celebration to inaugurate the new senior citizen housing project. On a lighter tone it is suggested that all the men in Centerville grow beards.[8]

The single building contributes to this scenario; activities imbue the places with significance, suggesting that we should never look at structures independently of their occupants and surroundings.

For the architectural historian the liberties taken with "facts" and the creation of "fiction" may be a complete anathema. And one can rightly question the importance of such forms in a field dedicated to "scholarship" and "objective" study. First, we must recall that Jackson was not an architectural historian, nor did he write (primarily) for an academic audience. But that is not to deny the importance of his work as a means to see a broader picture and to realize that life—and architecture—is a far more complex network of intersecting and interlocking factors, both animate and inanimate. His approach provides a broad base and a frame for understanding; it never pretends to offer the detailed examination of extended study. Neither qualification, however, denigrates the significance of his approach to those concerned with the history of the environment.

I first met Brinck in the late 1970s; he had begun teaching at Berkeley in 1962 at the invitation of the geographer Carl Sauer. Due to the intervention of landscape architecture chair Leland "Punk" Vaughan, however, Jackson's home base became the Department of Landscape Architecture, where he taught each winter term from 1967 for about a decade. "In the end, my topic was named 'The History of the American Cultural Landscape'—by which was meant the natural environment as modified by man."[9] His course was a university-wide favorite, consistently drawing a packed room of 300 students.[10] In a mild confession, Jackson later recounted:

> The more I sought some justification for discussing the cultural landscape at such length, the more convinced I was that the course had little practical or scholarly value. So my contribution to the education of my students was simply this: I taught them to be alert and enthusiastic tourists.[11]

Tourists? "An embarrassing kind of insight!...I was...aware of the fact that what I had tried to share with students was precisely the pleasure and inspiration I myself had acquired not from books, not from college, but from many years of travel. What I was passing on were those experiences as a tourist—or the means of acquiring them —that had been most precious to me."[12] Jackson told us all to begin by looking not at books, but at the world around us, for here is where learning about the environment begins. And that is how he influenced the vision of a generation of designers.

His course began and ended with the American environment. I sat in on his lectures for the first time in the late 1970s, at a time when my work involved the vernacular environment of signing, shaped buildings, and graphic communication. I was moved by the concision of Jackson's

ideas, and the simplicity and ease of his delivery. Jackson's method contrasted markedly with that of Spiro Kostof, another colleague and friend. Spiro's lectures were packed to the brim with ideas and facts, delivered with bombast and fluidity as if from an evangelist's tongue. Jackson, in contrast, was a softly spoken avuncular commentator, engaging his audience as if in a fireside chat. Typically, he spoke for forty minutes and then showed slides for another ten, reiterating the points previously made. The style of the architecture illustrated meant little to him; more important were the generalizable ideas about structures and their agglomeration as farms, towns, and cities. He used but one projector and showed slides of dubious exposure and framing, at times with a telephone pole squat in the middle of the view, at other times as if shot from a still-moving vehicle.

Perhaps the most gratifying aspect of his lectures was the elevation one felt by his citation of something quite commonplace and known; one actually felt proud of some aspect of the American environment, however insignificant. Perhaps it was his describing how the ecology of the roadside differed markedly from that of the surrounding fields or wood, or why our factories and office parks took the forms they did. While never an American chauvinist, Jackson raised our consciousness of what we have achieved as a people in terms of our living environment. He probably would have bristled as much about the demands of un-realistic diversity as he did at the environmentalists who wanted to freeze landscapes in time, or those sentimentalized notions of wilderness proffered by the Sierra Club. The former hope to construct difference, the latter to hold at bay those processes that must accompany the passage of time.

Instead, Jackson looked at what we share, what we have in common. He had the ability to strike up a conversation—a mild, seemingly benign, interrogation actually—with virtually anyone. A year or so ago we took a small excursion to have lunch in Chimayo, New Mexico, and on our way out of the restaurant encountered a family celebrating some sort of anniversary. Brinck was having a smoke and asked them why they had come. Within fifteen minutes he knew of the family, the significance of the occasion, why they had chosen this restaurant to celebrate the event, and the mutual perspectives of adults and children alike. And he remembered what they had said, and wanted to discuss it with us in the car on our return to La Cienega because they were significant to him [9-3].[13] Events like these became the stuff of his ideas and his essays. In 1994, for example, he wrote of a night auto mechanic course he took while a visiting professor at the University of Texas at Austin, "with a very easy schedule and a great deal of spare time."[14]

John Brinckerhoff Jackson was a great man with enormous accomplishments, whose gifts to humanity were as much by deed as by word. While living in an unassuming manner in La Cienega, he functioned as the *padrone* of the village. He quietly lent money to people whom he knew would never return it. He built a community center and pool for the village and when the latter fell into disrepair, he paid for its restoration. In visits to his house, one encountered a stream of people —almost always in the kitchen, where life truly centered—asking for a loan, bringing him cakes or meals, coming to thank him for putting a young man through college, even inquiring whether he would like to purchase a television or some other consumer good of uncertain ownership [9-4]. One rarely heard him say no to these requests (except, of course, to the offer to buy those goods of uncertain ownership). He was a true philanthropist, probably giving away in the afternoon what he had earned in the morning.

Jackson's project remains a vital model for its very breadth, its regard for every life situation as a part of something greater, and for the courage to define the landscape in other than pictorial and formal terms. Late in life, he wrote:

> The older I grow and the longer I look at landscapes and seek to understand them, the more convinced I am that their beauty is not simply an aspect but their very essence and that that beauty derives from the human presence... The beauty that we see in the vernacular landscape is the image of our common humanity: hard work, stubborn hope, and mutual forbearance striving to be love. I believe that a landscape which makes these qualities manifest is one that can be called beautiful.[15]

I also very much admire the clarity with which he wrote and his ability to communicate his ideas, the breadth of his knowledge and vision, and his approach to each subject of study free of prejudice, pursued with a fresh eye and strategy. But perhaps most important, Brinck Jackson demonstrated to me the difference between knowledge and wisdom. To be knowledgeable is more easily accomplished; the gathering of facts is like the raising and harvesting of crops. But this is fodder, it is the raw material that must be integrated with thought, ideas, and experience. It needs to be tested and tempered, distilled and clarified. Thus comes wisdom. Perhaps, I am a bit too much in awe of the person, too close to his wonderful personality, to ever be able to evaluate his work objectively. I am more proud of his dedication of *A Sense of Place, A Sense of Time* to Dorothée Imbert and me than I am of anything I myself have ever written. When John Brinckerhoff Jackson died this past August, we all lost a part of our intelligence, our sensibility, a part of our vision and voice. We have benefited greatly from his ideas and his writings, and may continue to benefit from his manner of studying the cultural landscape should we choose. I certainly will; and I would very much like to be like him when I grow up.

Originally published in the *Journal of the Society of Architectural Historians*, December 1996.

I wish to thank Nicholas Adams, then editor of the journal, for soliciting this essay.

Notes

1 John Brinckerhoff Jackson, "Landscape as Seen by the Military," in *Defining the Vernacular Landscape*, New Haven, CT: Yale University Press, 1984, p. 133.

2 Ibid., p. 135.

3 Ibid., p. 136.

4 The most comprehensive collection of Jackson's essays is Helen Horowitz, ed., *Landscape in Sight: Looking at America*, New Haven, CT: Yale University Press, 1997.

5 John Brinckerhoff Jackson, "Other Directed Houses," in *Landscapes*, Amherst, Mass.: University of Massachusetts Press, 1970, pp. 56–57.

6 John Brinckerhoff Jackson, "The Westward-moving House," in *Landscapes*, p. 31.

7 John Brinckerhoff Jackson, "The Necessity for Ruins," in *The Necessity for Ruins*, Amherst, Mass.: University of Massachusetts Press, 1980, p. 90.

8 Ibid., p. 97.

9 John Brinckerhoff Jackson, "Learning about Landscapes," in *The Necessity for Ruins*, p. 2.

10 John Stilgoe served as Jackson's teaching assistant at Harvard, where he taught each fall semester, and formalized many of Jackson's lessons in John Stilgoe, *Common Landscape of America, 1580–1845*, New Haven, CT: Yale University Press, 1983.

11 John Brinckerhoff Jackson, "Learning about Landscapes," in *The Necessity for Ruins*, p. 3.

12 Ibid.

13 For an overview of Jackson's landscape, see Marc Treib, "J. B. Jackson's Home Ground," *Landscape Architecture*, Volume 78, Number 3, April–May, 1988.

14 John Brinckerhoff Jackson, "Looking into Automobiles," in *A Sense of Place, a Sense of Time*, New Haven, CT: Yale University Press, 1994, p. 168.

15 John Brinckerhoff Jackson, "The Word Itself," in *Defining the Vernacular Landscape*, p. xii.

[10]

Looking Forward to Nature:
Garrett Eckbo, an Appreciation

2000

The completion of Garrett Eckbo's Alcoa Forecast garden in 1959 created quite a stir [10-1]. Executed in various shapes and meshes of aluminum, anodized in glittering metallic tints, the garden's pergolas and screens smacked of ultramodernity and an optimism for the future. The war had ended over a decade before, and the transition to a peacetime economy had been successful. Service personnel had returned to the United States to work or to study, and to regain the normalcy of daily life. That aluminum had made the jump from the defense industries to the garden was in itself significant, a selling point made in the widespread coverage the Alcoa Forecast garden received in the popular as well as the professional press.[1]

With Thomas Church's 1948 Donnell Garden and Dan Kiley's 1955 Miller Garden, the Alcoa garden can share ranking as one of the three most significant modern American gardens. While certain aspects of its overall planning were less striking in their novelty, there was no denying the invention expressed by the new material, or the permeable spaces and dappled shadow patterns the screens and roofs produced. In this one project, perhaps, we can witness a condensation of what Garrett Eckbo had argued for from his completion of graduate studies: a landscape architecture of its times that employed spatial design and contemporary materials to create appropriate settings for human activity.

Eckbo was born in Cooperstown, New York, in 1910, but his mother moved to Alameda, California, two years thereafter, and he with her. He studied landscape architecture at the University of California and graduated in the midst of the great depression. Through one of his professors, he got work in southern California in the Armstrong Nurseries and planned out about a hundred gardens during the year he remained there. These designs reflected the relaxed formality of their time and place, heavily planted with shrubs and fruit trees befitting designs commissioned from a nursery. They were hesitant in departing from the accepted norm and yet peppered with singular details that suggested greater creative explorations.

While satisfied to some degree with his designs, he sought graduate study at Harvard University to explore new paths and ideas and to see the other side of the country. His winning entry for a national scholarship competition provided the ticket. Eckbo's Harvard years would be a key period for developing his comprehensive views of landscape architecture and for realizing the social role for the profession. For one, his contact with classmates James Rose and Dan Kiley prodded

[10-1]
ALCOA FORECAST
GARDEN.
LOS ANGELES,
CALIFORNIA. 1958.
GARRETT ECKBO.

his own emerging sense of modernity in landscape design. With them, too, he shared an opposition to the school's conservative powers that were. As a result, he found more solace and interest in the architecture program's offerings than in those of the landscape department. Most important perhaps, his contact with the German émigré architect Walter Gropius encouraged a nascent social involvement, instigated by Eckbo's own modest, working-class background. For the remainder of his life, he would always see things from a definitely left-hand perspective, even though his practice would serve the private elite as well as the general populace.

After graduation in 1938 he worked briefly for Norman Bel Geddes on the landscape for the General Motors Pavilion for the 1939 World's Fair. His schemes—still preserved—tested varying roles for landscape in relation to a modernistic building.[2] None of his proposals was used; Eckbo believed that Bel Geddes wanted nothing that might detract from the sculptural forms of the pavilion. Through a crucial contact with Frederick Gutheim, he landed a job in Washington with the United States Housing Authority. In the six months he remained there, Eckbo began to address the question of spaces for modern apartment living aimed at the less fortunate segments of the population. No matter how restricted these might have been, they were almost luxurious in comparison to his next task: the design of migrant workers' camps in California, Arizona, and Texas.

Lured back to California with a position in the Farm Security Admin-istration, Eckbo felt he had found his niche. Here was important work, set within the broad spaces of California's Central Valley, created under the New Deal as the minimal vestige of human decency in housing agricultural workers.[3] The constraints were stringent, and the projects were executed with the most minimal of means. The ideas, in contrast, were high blown, employing sophisticated spatial models derived, in places, from architectural modernism such as Mies van der Rohe's 1929 Barcelona Pavilion. The key requirements were protection from the wind and blowing soil, shade from the grueling sun, and most of all, a landscape to integrate the diverse elements of the camp into a sanctuary from work, heat, and exploitation [10-2]. These were the camps so celebrated in John Steinbeck's *The Grapes of Wrath* as the only havens for the migrant workers, and Eckbo would remain most proud of his contributions for the remainder of his life.[4]

The western division of the FSA—which included architects Vernon DeMars and Burton Cairns, and landscape architect Francis Violich —was condemned as creeping socialism yet grudgingly lauded as a unit that could get things done. When the United States entered the war in 1941, the agency's efforts shifted from housing farm workers to housing the defense workers that were pouring into the Golden State to man the munitions, aircraft, and materiel plants. The Bay Area's design talent produced a host of laudable projects for the program, for example, William Wurster's housing in Vallejo (with Thomas Church as the landscape architect) and Eckbo's own contributions to several community plans.[5]

With the hostilities over, however, a basic optimism prevailed. Many of those who had come to California for work, stayed for the opportunities offered; returning GIs swelled the population, married, and procreated. Suburban enclaves sprouted around the major cities, and Los Angeles began to assume the sprawling character for which it would become famous. With Robert Royston and Edward Williams, Eckbo in 1945 founded a firm that would address the larger as well as smaller issues facing landscape architecture at midcentury. Their projects, almost all of a quality acknowledged by their peers, ranged from the small garden to the college campus, from the urban plaza to the subdivision and town center. In contrast to Thomas Church, for example, whose work tended to begin with a consideration of the smaller unit and then work centrifugally toward its greater context, Eckbo, Royston and Williams began with a more comprehensive vision that acknowledged the neighborhood and the district, and at times the region.[6]

Sensing that great opportunity lay in the burgeoning economy of the Southland, Garrett and Arline Eckbo moved to Los Angeles in 1946, establishing there a southern branch of Eckbo, Royston and Williams.

Of the range of landscapes produced by the office, it was the garden for which Eckbo became first known. The patterns of Californian outdoor living, paired with money from the postwar economic bounce, fueled the development of designs that smacked of contemporaneity. Centering on client desire, many of these designs featured extensive paved areas destined for entertaining while reducing the need for extensive gardening and maintenance. Space rather than form was Eckbo's target. As he put it: we live in spaces not on flat areas, however beautifully composed. His own compositions drew on contemporary art, the paintings of Joan Miró or Wassily Kandinsky, the architecture of Mies van der Rohe or Richard Neutra, the sculpture of Isamu Noguchi or Henry Moore. Projects such as the unrealized Burden garden in Westchester County, New York, of 1945 condensed many of these influences into a single composition, with forms and space, architecture and landscape entangled into a single residential and landscape knot [10-3].

The Eckbo gardens numbered in the high hundreds if not over a thousand. Some, like the 1948 Goldstone garden in Beverly Hills, began with the confines of the site and looked inward. Others, such as the 1956 Edmunds garden in Pacific Palisades opened beyond the terrace to exploit a distant and dramatic view. Few landscape architects in the Southland spoke to clients and their architects in just the way that Garrett Eckbo did, and as a result, there was a continuous stream of commissions, in many ways, each feeding the next.

Eckbo's ability to derive impressive formal innovation from client need is well illustrated by two swimming pools accompanying garden designs from the mid-1950s. The Cole garden in Beverly Hills answered the call of the noted swimwear designer for a setting equally suitable for announcing the season's fashions as well as for the quotidian dip. The wall at the rear of the pool appears to be its far limit, yet it spans the pool supported on a concrete beam, allowing the water to continue beneath [10-4]. One can imagine the excitement when swimsuit models entered the pool—screened from view by the masonry wing walls— and emered Esther Williams-style, wet and golden.

The Cranston pool developed from an entirely different scenario. Rather than display, Eckbo here addressed the practice of physical

[10-3]
BURDEN GARDEN.
WESTCHESTER
COUNTY, NEW YORK.
1945.
GARRETT ECKBO.
Plan.

[10-4]
COLE GARDEN.
LOS ANGELES,
CALIFORNIA.
EARLY 1950s.
GARRETT ECKBO.

[10-5]
CRANSTON GARDEN.
LOS ANGELES,
CALIFORNIA.
EARLY 1950s.
GARRETT ECKBO.

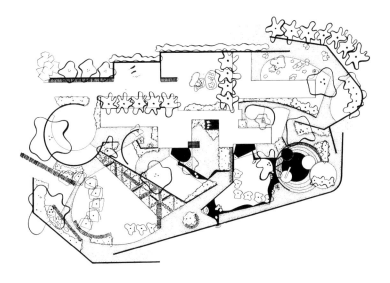

rehabilitation for a paraplegic client. The curious lambchop shape derived from an elongated, shallow ramp that allowed the owner to enter the pool by wheelchair under his own power; the positioning of the small island provided a place for upper body development [10-5]. We see here, once again, Eckbo's impressive ability to derive spatial and formal interest from seemingly restrictive needs.

By the later 1940s there had collected a sufficient body of work, and a gathering intellectual interest, to consider publishing ideas drawn from professional practice, observation, and thought. *Landscape for Living* appeared in 1950, a tract that in many ways has never been surpassed in terms of the quality of the work illustrated and the comprehensiveness of its vision. In the course of the century remarkably few books had treated the subject of modern landscape, and only one of them—Christopher Tunnard's influential 1938 *Gardens in the Modern Landscape* (or more precisely, its 1948 revised edition) had attempted to outline a theoretical stance for landscape practice in the later twentieth century.

The illustrations in *Landscape for Living* were drawn primarily from the work of Eckbo, Royston and Williams, and in that sense the book was as much an apologia for the firm as a theoretical tract. But the text, in contrast, assembled ideas from an impressive panoply of sources, replete with readings from psychology, that would become ecology, sociology, and even political philosophy. Its dense writing does not make for easy reading and it demands much from the reader, even today. One would imagine, in fact, that most landscape designers bought the book for its images and few ever took the time to read the text, which was in part manifesto, part plea for sanity in developing the land. Ian McHarg's *Design with Nature* of 1969 has achieved a broader audience, both professional and popular; but *Landscape for Living* provided a more lucid model of how platitudes can find realization in landscapes of exceptional innovation and quality.

In time, there were other books as well. *The Art of Home Landscaping* of 1956 addressed the popular audience almost as *Sunset* books might today. The main concern—illustrated in a series of witty cartoons at the front of the book—was comprehensive site planning. For Eckbo, thinking should always precede action; the more the thinking up front, the less the reworking later—we see the consequences of acting too quickly in the last cartoon in which the family that jumped first is paying for their sins of hasty decisions. *Urban Landscape Design* broadened the arena to the public sphere, featuring such projects as the landmark Fulton Street Mall in Fresno, California (designed with Victor Gruen), one of the first conversions of a city street into a pedestrian zone [10-6].[7] The work for Ambassador College in Pasadena and England, Orange County College, and the zippy curving forms of the Union Bank suggested Eckbo models for landscapes larger than the garden. *The Landscape We See* of 1969 took a different tack, laying out a more theoretical, although highly empirical, conception of our reading of the world around us. In some ways the book was an update and broadening of *Landscape for Living*, with relatively few images set within the text and nearly nonstop sidebar quotations.[8] More a philosophy than a portfolio, the book proposed "total landscape," a concept by which Eckbo sought "quality" in the relationship between human and nature.

Art was one of the principal means by which this quality would be effected. "Art is, among other things, the creative portion of the processes of change," Eckbo wrote.[9]

The expansion of view expressed in the writings paralleled an expansion of concerns and scale in professional practice. It became obvious to Eckbo that although suburbia remained an ideal living environment for the postwar Angeleno, the natural environment would suffer severe repercussions as a result. To some degree accepting of the thriving settlement trend, he undertook a number of projects in which he attempted to realize a suburban landscape that would strengthen community identity and quality while allowing high degrees of individual expression. Eckbo's use of varying species of trees at the Mar Vista houses, for example, complemented the more or less standard, though flexible, houses by Gregory Ain. Over time, the various owners would modify the stock unit and would plant their own front and rear gardens. Despite this increased individuality, as the street trees grew the coherence of the community would also be strengthened.

Eckbo's most coherent vision for suburbia remained unbuilt in Reseda, in the San Fernando Valley [10-7]. Community Homes was planned in 1945–1949 as a racially mixed community for the Screen Workers Guild. On an ell-shaped piece of land, the site was platted for house variants, again designed by Gregory Ain. The street plan allowed children (and adults) to pass to schools and parks by crossing only a minimum traffic. The tree planting plan furthered ideas first implemented at Mar Vista, but here Eckbo used species of particular shape and volume—palm trees versus eucalyptus, for example—to shape superscaled zones, almost as a gigantic version of the flowing spaces of Mies van der Rohe's Barcelona Pavilion. Interestingly, he developed his own set of diagrams to design and track the space-defining qualities of each of these masses. Sadly, due to federal loan restrictions, the project remained stillborn. Its broad ideas and specific proposals still remain valid, however, and wait to be rediscovered and implemented as a contemporary alternative to many of the *retardataire* aspects of the so-called New Urbanism.

The design projects grew into planning projects, both for the community and the region. And the firm metamorphosed as a consequence. Eckbo, Royston and Williams was dissolved in 1958. Francis Dean had become

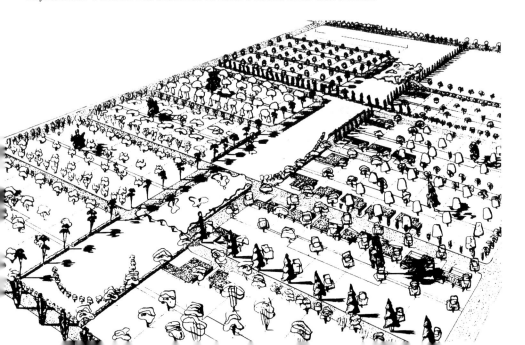

a partner in 1948; the firm Eckbo Dean Austin and Williams was incorporated in 1964 (still extant today, in a different and greatly expanded form as EDAW). The focus of the firm's work shifted from detailed design and the garden to larger and larger planning projects, including Eckbo's own participation on the committee to tackle the problem of the deterioration of Niagara Falls. While the design aspects of the later work were less apparent and less inventive as formal exercises, the high quality of the thinking remained constant.

In 1963, Garrett Eckbo accepted the call by his alma mater to return to the Bay Area to chair the Department of Landscape Architecture at the University of California at Berkeley. He remained there over a decade as chair or as professor, receiving emeritus status in 1978. In his teaching, as in practice, he had a quiet way, accepting of diverse viewpoints, slow to condemn ideas out of accord with his own. He remained a gentle personality to the very end [10-8]. He paired up professionally with Kenneth Kay in 1983 and continued to work on residential and civic projects; by that time the ecological wave had rolled over most interest in formal innovation, and Eckbo devoted much of his time studying possible antidotes for thoughtless development and the thoughtless extraction and use of natural resources. In later years, he worked almost daily on his writing, and a last collection of diverse thoughts appeared in 1998 as *People in a Landscape*.[10]

How does one evaluate the influence of Garrett Eckbo and measure his contribution to the profession of landscape architecture and the American environment? To me personally, Eckbo's greatest lesson will be found in the sweep of his thinking, his tackling of ideas as abstract as "science" and "progress" and as specific as a paving material, a light quality, or the fixation on a particular demand of the program. The designs, seen almost a half century later, are stunning: always looking forward rather than backward, never with a hint of nostalgia. He was a man who looked always to the arts, for like the sciences, they reflected the accomplishments of our times. And he was a man always concerned with the human being taken as a specific client, the communal body, or what we would call today a user group. Abstraction never superseded the role of the individual in his designs.

Few of his landscapes remain unscathed. There seems to be an unwritten law that the more specifically a landscape reflects its times, the faster it appears dated, and the greater its difficulty in surviving until it receives the hallowed status of classic work. The Alcoa garden disappeared about two decades after its making, shortly after the Eckbos moved north to Berkeley. In 1996 the framework of several pergolas still existed, as did the experiments with ceramic pipes used as planters, but the aluminum magic was long gone and a naturalistic swimming pool occupied the garden's central area.[11] Other gardens retained their hardscapes—the structures, the pools, and the paving —but their planting had disappeared, or had been replaced, or had matured unchecked to the point that it no longer reflected the original design intentions. No doubt, in time, house owners will contact the Environmental Design Archives at the University of California, Berkeley, where the Eckbo papers are held. They will want to recreate the gardens not only as a retro period piece, but as a setting that reflected the best landscape design thinking of its time.

In *The Landscape We See*, Eckbo cites the painter Adolph Gottlieb: "Certain people always say that we should go back to nature. I notice they never say we should go forward to nature. It seems to me they are more concerned that we should go back, than about nature."[12] Garrett Eckbo was always that exception, a landscape architect who always looked *forward* to nature.

How can we characterize the landscape architect Garrett Eckbo? Humanist, thinker, and artist.

Originally published in *Landscape Architecture*, December 2000.

Notes

1 *See Marc Treib, ed.,* Thomas Church, Landscape Architect: Designing a Modern California Landscape*, San Francisco: William Stout Publishers, 2004.*

2 The Eckbo papers are held at the Environmental Design Archives, College of Environmental Design, University of California, Berkeley.

3 In retrospect, Eckbo wrote: "Innovative site planning, engineering, and design made the most of these opportunities by creating memorable spatial experiences. The high value of shade in these hot interior areas guaranteed maintenance and irrigation for the trees. Responsible management guaranteed safety, security, and social relaxation within the projects." "Pilgrim's Progress," in Marc Treib, ed., *Modern Landscape Architecture: A Critical Review,* Cambridge, Mass.: MIT Press, 1993, p. 218.

4 John Steinbeck, *Grapes of Wrath,* New York: Vintage, 1939.

5 For Wurster and Church's col-laborative projects, see Dorothée Imbert, "Of Gardens and Houses as Places to Live," in Marc Treib, ed., *An Everyday Modernism: The Houses of William Wurster,* Berkeley, CA: University of California Press, 1997, pp. 114–137.

6 Edward Williams, *Open Space: The Choices before California,* San Francisco: Diablo Press, 1969.

7 Garrett Eckbo, *Urban Landscape Design,* New York: McGraw-Hill, 1964.

8 Garrett Eckbo, *The Landscape We See,* New York: McGraw-Hill, 1969.

9 Ibid., p. ix.

10 Garrett Eckbo, Chip Sullivan, Walter Hood, and Laura Lawson, *People in a Landscape,* Upper Saddle River, NJ: Prentice-Hall Inc., 1998.

11 *The garden is extensively discussed and illustrated in Marc Treib,* The Donnell and Eckbo: Gardens: Modern Californian Masterworks,*San Francisco: William Stout Publishers, 2005.*

12 Eckbo, *The Landscape We Se*e, p. 35; quotation from *The New York School,* Los Angeles County Museum of Art, 1965.

[10.8]
GARRETT ECKBO WITH
SYLVIA CROWE.
BERKELEY,
CALIFORNIA. 1982.

[11]

Postulating a Post-Modern Landscape

1985

Every era has referred to its age as modern; ours is no exception. "Contemporary" signifies of our period, of our times [11-1]. Over the last century the prevailing course of art history, termed "modern art," progressed toward and ultimately retreated from the pursuit of expressing its own being. Promulgated by such art critics as Clement Greenberg, this theoretical stance saw painting first and foremost as a series of significant marks made on canvas.[1] Since the implementation of perspective in the fifteenth century until well into the last century, painting sought to capture time and place—as a representation of a real or imagined existence. But need that be its necessary purpose?

Impressionism made the first major breach into the bastion of representation, although the shelling had begun earlier in the nineteenth century in the works of painters such as Gustave Courbet. An emphasis on the act and manner of seeing and painting itself replaced pictorial clarity, shifting to some degree the focus from that painted to the manipulations of the painter. Monet linked painting with light, replacing reality as measured with reality as perceived. Early in our century, the cubists shattered the coherent view of space and in its place delivered a composition of multiple views in space or in time. In so doing painting was drawn, they believed, closer to "reality."[2]

In the post-World War II era painting continued to seek its essence. Ad Reinhardt's saturated black and blue canvases, Yves Klein's indigo paintings and Jackson Pollock's overall calligraphic textures collectively advanced the position of the artist's action or artifact as critical to the aesthetic presence—and ultimately the significance—of the work. By the 1960s, the primary or "dumb" unity of minimalist sculpture replaced gesture with a concern for focused perception in real time and real space. But even that vestigial existence evaporated with the advent of a conceptual art that reclaimed certain foundations laid by Marcel Duchamp earlier in the century. Duchamp saw art to be largely a matter of intention and believed that content resided in the thought or its record and not in the artifact itself. Within the purview of such a philosophical stance, the world of reality and the world of art could be regarded as coincident.[3]

Architecture shared certain formal properties of painting and sculpture; though socially burdened, its development followed at a far slower pace than the artistic mainstream. The fulfillment of the twentieth-century program for air, light, greenery, and sanitation was hampered by the retention of traditional notions of form such as the adherence to the vocabulary of historical and classical styles. One must remove these structures of meaning, the modernists argued, to derive meaning, instead, from structure. Painting had been redefined as placing marks upon a canvas. Architecture, in a parallel manner, could be thought of as the act of building from disparate elements for social purpose. Modernist architecture, then, would draw upon the system by which it was produced, utilizing its architectonic properties as the basis of its idiom. Formally speaking, space was the content, production the form.

Landscape architecture lagged farther behind architecture in concerning itself with those aspects of aesthetic development explored in the fine arts. Well into this century the landscape mentality was governed by the grand-scale picturesque that drew upon the naturalistic gardens of Capability Brown and Humphry Repton and the parks of Frederick Law Olmsted, Sr. America was rapidly becoming small, its wilderness circumscribed by development. With the threat of the unfamiliar removed by increased settlement and the productive utilization of land, the wilderness could be appreciated and appropriated, with attempts made at its replication in however reduced a scale.[4]

The occasional forays into the classical landscape vocabulary virtually completed the palette. Formal gardens, with their control and elaborate upkeep, were traditionally the domain of the rich, who possessed the resources to create and maintain them. The accompanying references to the stability and formal tradition (and aristocracies) of Europe were also well taken.

The ecological movement that matured in the 1960s injected new credence into this flagging tradition. Ian McHarg's *Design with Nature* outlined a comprehensive approach to regional landscape planning but suggested little in aesthetic terms except the pattern of minimally disturbed wild areas or a de facto extension of the neo-picturesque.[5] In California, landscape architects took up the call and placed the emphasis on the "natural" (whatever that might be) — even three trees in a row could be regarded as formalistic heresy.

But certainly there had been a modern innovative tradition in landscape architecture. The work of Garrett Eckbo, Thomas Church, and Dan Kiley peppered the periodicals of the times — professional and popular alike — in addition to works by the Brazilian Roberto Burle Marx which had been known in the United States from the 1930s. Their landscapes were of our time, hence contemporary by definition, and looked to be of our time in their formal vocabularies. But were they truly modern in the same sense of painting or even architecture of the period? My personal view is to think not—or let us say, not completely.

If one used the formal aspects of modernist architecture as a comparison, by extension a modern landscape should also express its own production. This logic could lead, for example, to an exceptional fondness for specimen plants; or, at an extreme, the botanical garden itself. Or perhaps the garden might feature the plowed furrow as a design form or use the irrigation system as part of the composition. Materials and production, as in architecture, would be converted to aesthetic appliances.

Many of the gardens by Burle Marx or Church utilized the forms of modern painting and design—great swatches of vivid color in gently meandering curves, for example—but these exploit only the hue of the plant, perhaps its texture [11-2]. In scale of application and in vocabulary, these features are certainly "new," but they are, in sum, more an extension of the grand parterre tradition than an exploration of the spatial developments of modernism.

Certainly there were exceptions. One can find, for example, the necessary contouring or terracing of slopes to thwart erosion used for aesthetic effect. Isamu Noguchi's use of sprinklers and other fountain devices at the 1970 Osaka Expo integrated sound and spray and refraction to further the sensual dimensions of the static pool. But for the most part these were departures rather than the norm; the accepted landscape trends pursued the idea of "settings for living:" comfortable, semi-formal gardens to complement the emerging modern California house. It was here, perhaps, that landscape architecture best complemented contemporary building. By effectively extending the interpenetration of inside and out, the garden furthered the spatial bond so endemic to modern architecture.

To date we have seen relatively little substantive thought toward formulating a post-modern landscape architecture. From time to time an article appears in the professional journals on some pathetic attempt at decorating a space in an unusual material such as cut-out Plexiglas [11-3] or at its most infamous: with bagels. That these minute projects should invoke such a vehement reaction from practitioners underscores the limited theoretical state in which the profession currently exists. While these works may provoke our curiosity, they often lack

[11-2]
CAVANELLAS GARDEN.
PEDRO DE RIO,
BRAZIL. 1954.
ROBERTO BURLE MARX.

[11-3]
KITAGATA HOUSING
LANDSCAPE.
GIFU, JAPAN. 1999.
MARTHA SCHWARTZ.

that critical feeling of essence or substance that endows the world's more significant landscapes.

Ironically perhaps, it is the architects who have advanced many of the most interesting current landscape ideas, extending their domain to include once again the garden or the greater limits of the site. There are at least two reasons for this. First, architecture has recently "rediscovered" the beauty of classical composition and is unafraid of applying its virtues to the surrounding garden. Second, reference to historical precedent has been taken by some as the means to grant an air of substance and significance to new construction. However naïve this thought might be, it has opened—or reopened—a wealth of historical landscape sources to possible interpretation denied by a previous generation.

Post-modernism in architecture does not yet possess a commonly shared definition—nor does the art world agree on one either. Some dismiss the term as meaningless, indicating chronological position rather than aesthetic development per se. Others regard post-modernism as a furthering or an improvement on the accomplishments of modernism, whose sources of meaning—construction and tectonics—they regard as solely syntactic rather than truly semantic systems. Only in a return to traditional semantic sources—such as type or historical style—can we once again regain a broadly meaningful expression, they maintain.

At the extreme, proponents claim that only by employing the classical language of architecture can a meaningful architecture result.[6] Classicism is the language of tradition, of continuity. Its forms are known; its values assured. On the other hand, those who value type over style dismiss the classicist stance as hogwash. It isn't the clothing, they rejoin, but the disposition of the body that is critical to signification.[7] Yet another group seeks a more immediate source of inspiration in the immediate surroundings, deriving an architecture in response to these conditions and approach, characterized as contextualism.[8]

What are we to make of these approaches? What could we extract or extrapolate from them toward postulating a post-modern landscape?

The argument for reinstating a classical landscape, such as the formal gardens of the Italian Renaissance or the French Rococo periods, seems wide of the mark. These gardens, as styles or types, shared little of the widespread application that marked classicism in architecture.

Like the recent re-introduction of architectural neo-classicism, the implementation of diluted styles appears anemic at best and ludicrous at worst. Does type, then, provide a direction for landscape architecture? Potentially yes, I might answer (it does seem like a compelling idea), though I would be hard put to advance a formulation of gardens or garden element types that would be sufficiently specific to be useful in designing today. This leaves us with the notion of context and the residual unfulfilled project of modernism.

Let us take the latter first. Earlier I suggested that the true modernist landscape had never fully developed and that landscape architects had tended to apply only a vocabulary drawn from modernist painting to designs for planting and paving. Certainly the complexity of cubist spatial conception would be an excellent starting point for future developments. While small gardens such as those on the rue Mallet-Stevens in Paris, designed by the architect himself, were executed in a cubistic style, using a cubist vocabulary, their layouts read suspiciously traditional [11-4].[9] What would a landscape of overlapping and simultaneous views suggest in terms of an aesthetic? In certain ways the controlled composition of the English picturesque gardens answers some of the criteria, but its sequence was restricted to a path (or paths) that was, in most instances, neither variable to any great degree, nor reversible [11-5]. What of a garden that consciously commanded the intersection of routes—like Jorge Luis Borges's "Garden of Forking Paths"—that led and returned to vistas and visual termini?

[11-4]
VILLA NOAILLES.
HYÈRES, FRANCE.
1927. GABRIEL
GUEVREKIAN.

"Precisely," said Albert. "'The Garden of Forking Paths' is an enormous riddle, or parable, whose theme is time; this recondite cause prohibits its mention. To omit a word always, to resort to inept metaphors and obvious periphrases, is perhaps the most emphatic way of stressing it."[10]

What of a garden that used selected plant species so that vistas, objects, or structures were closed or revealed to view by the passing of the daily cycle, the seasons, the linear passage of time—or even the passing of clouds far overhead?[11] Such a garden would not only exemplify the cubist concept of space and presentation, but also the modernist reliance on the properties of materials as a basis of expression.

But, should cubism be used as a spatial model? Can an essentially flat medium provide a viable source of ideas for a four-dimensional environment? After all, the presentation (and composition) of simultaneity of alternate views or times was an attempt to replicate in two dimensions that which is inherent in architectural and landscape experience—the fourth dimension. The question is a good one—and I would have to say that the only defensible application of the painterly thought would be to enrich our normal experience of landscape rather than attempting to replicate a painterly movement in space—a new picturesque, so to speak.

Context alone remains for discussion, and it is context, I believe, that still offers us our greatest source of enriched landscapes. The concept is hardly new. It has probably existed in one form or another through all time, from the first efforts at agriculture in which the conditions of the place dictated the form of husbandry and determined the crops to be cultivated. Alexander Pope, of course, invoked the genius loci, the spirit of the place who should be consulted in all—and yet that consultation has only under the best circumstances led to designs truly of the site.[12]

One could say that Jens Jensen, who argued for the use of only native plant material also consulted the genius, but the forms that his planting took were often hyper-naturalistic interpretations of existing natural orders.[13] Rather, it should be the order of the place, its inherent structures as topography or vegetation, or its relation to surrounding architecture that ultimately dictates the conceptualization of form. The best examples of this school stem neither from architects or landscape architects today, but from sculptors.

During the late 1960s several sculptors, including Robert Smithson, Michael Heizer, Nancy Holt, Robert Morris, and Dennis Oppenheim executed a series of sculptural works upon or in the terrain. Usually isolated in the desert, where land was both abundant and cheap, these sculptors took the earth or its properties as the basis of the work. Heizer's *Double Negative* (1969–1970) is the clearest statement of order: two straight lines through the irregular perimeter of the mesa, bull-dozer cuts aligning across the eroded face of the cliff. Horizontal and vertical are mutually embedded as horizontal movement translates directly into vertical rise or fall. The material is entirely homogeneous and only the order in which the earth is formed distinguishes the "work" from the place. The material is both visible and invisible, manifest and dissolved, simultaneously. Ordering alone articulates its presence.

Walter de Maria's *Lightning Field* (1978) overlays the irregularity of the desert floor with a rectilinearly ordered field of stainless steel poles. Unlike the dichotomized monologue of earth to earth in the Heizer piece, de Maria here employs dialogue and definition through opposition. Thus order contrasts irregularity, the artificial (made) with the natural, and metric rhythm with the uncounted distance. But if we imagine a similar structure inserted, say, into a Parisian square—which in some ways is neither unthinkable nor uninteresting—we witness to what degree significance derives from the site conditions. Urban places, like natural terrain, possess their own systems of structure and instigate their own solutions in response.

Of course, the vast open spaces of the desert inherently possess their own power and any clever work taps this power rather than trying to devise its own. In a small space, an architecturally defined space, the power must be created. Isamu Noguchi's 1982 *California Scenario* garden at the Southcoast Plaza in Costa Mesa, California, establishes a set of oppositions and harmonies in accord with the sculptor's own abstract logic [11-6]. Environmentally, one can cite several problems, among them the lack of shade and the glare generated by the stark white walls in a climate characterized by heat and bright sun. One could also quarrel with certain particularities of the design and, in places, the order dissipates to the point that one feels that the space has been furnished rather than conceived as a whole. But formally speaking the garden is both a frontispiece and antidote to the inert mirror-glass boxes and the prosaic parking garage that surround it. Two grassed glades —one planted to wildflowers and redwoods—insert the High Sierra wilderness, however diminished, into suburbia. Stone appears as rough boulders or polished surface [11-7]. Tapping on our sense of the primitive and the essential are the stone pyramid and the wedge-shaped waterfall—recalling the eighteenth-century astronomical observatories of Jai Singh in Jaipur and New Delhi. A small water-course links the pieces, softly and tenuously. On the whole, despite its abstract conception, the garden seems to have responded to the conditions of the place—although differing significantly in order from the mono-tonous repetition of the curtain walls that reflect it. In this garden, the order of the landscape architecture and the understanding of the physical characteristics of the sculpture neatly intersect.

The design for the Columbus, Indiana, Carscape competition addressed the environmental memory of the site, recalling what had been lost in the course of the city's development. The layout of the parking lot is pragmatic and determined by the program for the required number of cars [11-8, 11-9]. On the perimeter of the two-block site are evoca-tions in steel mesh of the first five feet of the houses that were destroyed to provide space for parking. Planted with deciduous vines they would become, with the replanted front yards, the green ghosts of the neighborhood displaced by the new parking area. The earthen berms that buffer the subsections of parking roll in gentle slopes to recall the surrounding prairie and would be planted with native grasses, corn, or wheat. Thus, the design was intended to recall—simultaneously —three levels of the site's history: the natural; the agricultural; and the earlier period of habitation—each transformed for its own aesthetic purpose.

[11-6]
CALIFORNIA SCENARIO,
COSTA MESA,
CALIFORNIA. 1982.
ISAMU NOGUCHI.

[11-7]
*MONUMENT TO THE
LIMA BEAN*, AT
CALIFORNIA SCENARIO,
COSTA MESA,
CALIFORNIA. 1982.
ISAMU NOGUCHI.

If we consider significance to stem principally from perception rather than intention, it is perhaps the California landscape of the vineyard that comes closest to illustrating a true post-modern landscape [11-10]. While born of the fall of the land, the specifics of orientation, and the spacing for maximum yield, its play of regularity across the golden hills welds the cultivated to the indigenous landscape in a bond that simultaneously advances into presence yet recedes into coincidence. Though not plotted for aesthetic effect, it represents an understanding of order, fit, and use so often lacking in much contemporary landscape design. In postulating a post-modern landscape I can find no better descriptive qualities.[14]

Originally published in *Landscape Design: New Wave in California, Process Architecture*, Number 61, August, 1985.

[11-8]
"GREEN GHOSTS."
CARSCAPE
COMPETITION ENTRY.
COLUMBUS, INDIANA.
1982. MARC TREIB.
Plan.

[11-9]
"GREEN GHOSTS."
Lateral view of the
screens.

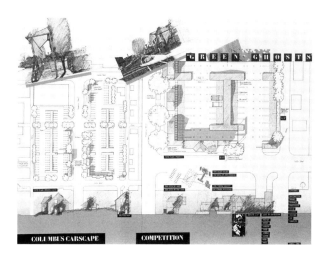

[11-10]
VINEYARDS.
YOUNTVILLE,
CALIFORNIA.

Notes

1 See Clement Greenberg, *Art and Culture*, Boston: Beacon Press, 1961.

2 See Robert Rosenblum, *Modern Painting and the Northern Romantic Tradition: Friedrich to Rothko*, New York: Harper & Row, 1975.

This breathless review of the course of painting over the last one and a half centuries has its obvious shortcomings, among them a lack of complexity, subtlety, and the acceptance of divergences from the canonic stream.

3 See Gregory Battcock, ed., *Minimal Art: A Critical Anthology*, New York: E. P. Dutton & Co., 1968, especially Robert Morris, "Notes on Sculpture," pp. 222–235. Also Ursula Meyer, *Conceptual Art*, New York: E. P. Dutton & Co., 1972.

4 See Nikolaus Pevsner, "The Genesis of the Picturesque," 1944, in *Studies in Art, Architecture and Design: Volume One*, London: Thames and Hudson. pp. 78–101.

This is obviously an oversimplification but the picturesque tradition in the United States shared little of the political or literary origins of the English version of the movement.

5 Ian McHarg, *Design with Nature*, Garden City, NY: Doubleday and Company, 1969.

6 Among them Alan Greenberg and, at times, Robert Stern.

7 Aldo Rossi is a good representative here. See his *The Architecture of the City*, 1966; English translation, Cambridge, Mass.: Opposition Books and MIT Press, 1982.

8 The early construction at Sea Ranch, California, well represents the contextualist school, particularly the cluster houses by Joseph Esherick and Associates (1965) and Condominium #1 by Moore, Lyndon, Turnbull and Whitaker (1965). Edward Larabee Barnes's Haystack Mountain School of Crafts (1960+) and Venturi and Rauch's Trubeck and Wislocki houses on Nantucket Island, Massachusetts (1970), demonstrate the ability to occupy the niche between both the high art and local vernacular traditions. See Charles Moore, Gerald Allen, and Donlyn Lyndon, *The Place of Houses*, New York: Holt, Rinehart and Winston, 1974.

9 *Robert Mallet-Stevens collaborated with Gabriel Guevrekian on the 1928 villa for Viscount de Noailles in Hyères, France. See Hubert Jeanneau and Dominique Deshoulieres, eds. See Rob Mallet-Stevens, Architecte, Brussels: Archives d'Architecture Moderne. 1980.*

This would be a far better illustration of a "cubistic" vocabulary applied to the garden. Some years after the publication of this article Dorothée Imbert published her definitive study of these works and the period: The Modernist Garden in France, *New Haven, CT: Yale University Press, 1993. She explains in great depth the strengths and the limitations of the landscape designers who worked in this manner—and the problems of applying vocabularies from one artistic genre to another.*

10 Jorge Luis Borges, "The Garden of Forking Paths," in *Labyrinths: Selected Stories and Other Writings*, New York: New Directions, 1964, p. 27.

11 In fact, a number of landscapes by Roberto Burle-Marx function in just this way.

12 Alexander Pope, "Epistle to Lord Burlington" (1731), in John Dixon Hunt and Peter Willis, eds., *The Genius of the Place: The English Landscape Garden 1620–1820*, New York: Harper and Row, 1975, pp. 211–214.

This subject was examined more fully in "Must Landscapes Mean? Approaches to Significance in Recent Landscape Architecture," Landscape Journal, *Volume 14, Number 1, Spring 1995, reprinted in this volume.*

13 The role of Jens Jensen in the Midwestern landscape is discussed in Catherine M. Howett, "Frank Lloyd Wright and American Residential Landscaping," *Landscape*, Volume 26, Number 1, 1982, pp. 33–40.

Also: Leonard Eaton, Landscape Artist in America: The Life and Work of Jens Jensen, *Chicago: University of Chicago Press, 1964; and Robert Grese,* Jens Jensen: Maker of Natural Parks and Gardens, *Baltimore, MD: Johns Hopkins Press, 1992.*

14 *Today I could propose other examples drawn from landscape architecture that would represent these basic premises, perhaps with more dimensions than those enumerated here. The relation of the vineyard landscape to designs by Peter Walker and others who use the grid, however, suggests that there was some merit in my original proposition. See Linda Jewell, ed.,* Peter Walker: Experiments in Gesture, Seriality and Flatness, *Cambridge, Mass.: Harvard Graduate School of Design, 1990.*

[12]
Settings and Stray Paths

1998

Over the last four decades, pressure from both social and economic forces has radically reconditioned the character of the design professions.[1] Designers now stress the breadth of their services, the inclusive grasp of their practice. This structural metamorphosis has been relatively rapid and conclusive, with large corporate offices emerging from the shaky foundations of the small studio practice. Both the organization and the goals of the profession have evolved to meet new challenges, many of them concerning clients, economics, and implementation, as well as the mode of practice.[2]

More a parallel than a direct result, landscape design as an aesthetic (as well as social) project closely allied with advanced ideas in painting and sculpture fell into disrepute in the years following the social upheavals of the late 1960s. At the community scale, citizen participation in the decision-making process attacked the authority of the designer or planning commission and sought to temper economic gain with civic well-being. The architecture profession broadened its services, again embracing the exterior as well as interior realms. Impulses from seemingly unrelated disciplines, and pressures to address a broader range of needs, also effected an influence on the way in which buildings were conceived and realized. In landscape architecture, the real threat to the planet as a whole and the "discovery" of ecology as a medicating procedure, turned the profession from its principal historical course: creating exterior settings of vegetal and inert material that understood natural conditions and patterns of human use. Ian McHarg's doomsday pronouncements in *Design with Nature* were instrumental in redirecting the attention of landscape architects from the aesthetic accomplishments of design—one could say their cultural and intellectual aspirations—toward a greater understanding of our position in the environment, both negative and positive.[3]

Analysis replaced formal consideration, as if the process by which landscapes were directed (if not created) could result from study alone. As in any era, the resulting designs varied in quality.

The broadening of the landscape vision, however, witnessed a diminution in the design's strength as perceivable space and form. Then came the backlash.

In the 1980s, a new generation of landscape architects saw in distant history (that is, periods greater than 20 years ago) a long story of accomplishment and beauty. Almost in one fell swoop, many of the lessons of the Ecological Age were lost, and designers instead turned for inspiration to artists and their works in the landscape.[4] This was a loaded view, however, since art—unlike design—begins with constraints internal to the artist him- or herself. In contrast, design remains a social practice (some might say social art) that needs to consider a far broader number of parameters. To be viable, landscape architecture must consider more than form or light quality alone, although this limited range of address is quite acceptable in the artwork. In this sense, artists can serve as the research wing of landscape architecture; design demonstrates applied development.

It is true that a number of landscape architects have been rewarded for being artists, and one can rightly develop definitions of practice that cross over traditional lines. A landscape architect can act like an artist, and an artist can assume the manner of a designer—categorizations depend on the approach and the project. This is not really the point. The question, it would seem, concerns the role of landscape architecture in a world continually losing an active engagement with the physical environment. In its place we seek a virtual reality, perhaps as an escape from, perhaps as an enrichment of, the world in which we actually dwell.[5]

The work submitted to the Designed Landscape Forum spanned a broad range of types and tread along rather stray, if well-trodden, paths. To speak of them as a group, to attempt to ascertain clear affinities and directions, is very much as if to compare apples and eggs. Other than their roughly spheroid volumes and food value, apples and eggs have little in common. The work we (the panelists) reviewed ranged from corporate landscapes for festive retail, to the private world of the residential garden, to the corporate campus, to the public park, to wannabe artworks. Only in their use of vegetation, and perhaps in the shaping of the ground plane, did they display common traits. Some asserted that ecological process must determine a naturalistic guise. Others suggested that to be a 1980s or 1990s landscape meant striking back with a vengeance at that accepted image of nature. Using patterns and order, they again brought nature under control, suspending the landscape in an eternal instant, circa 1989.[6]

The issue seems to be familiar. And although we would all prefer to lay to rest the question of the so-called constructed versus naturalistic manners, stylistically the submitted projects clearly exemplified just that very dichotomy. One group of designs appears to have left behind any hint of geometry for nature's greener pastures. Some appeared to advance nature as a model; others suggested that a constructed order, reflective of the human mind, is the more proper way to go. Obviously, the simple polarities of formal and informal do not, and perhaps can never, really exist. Instead, the particular choices of ordering and materials collectively exhibit a gradient of formalities— and with them varying attitudes toward what is "natural." A naturalistic landscape of plastic trees differs quite markedly from a similar setting using exotic vegetation, or another with all native species. Quite obviously, the prevalent order does not alone distinguish one work from another or establish its quality with any certainty.

Despite a great variety of particular idioms, some more natural, some more obviously treated as pattern or sculpture, landscape architecture at the end of the twentieth century seems to be falling into two major divisions. One group rejects any sense of identifiable form, exchanging that urge for the processes or look of nature. The second group rejects naturalism (its forms, if not its systemic performance) for an overt expression of human expression. (Yes, this is the old formal/ informal dichotomy once again, although here it informs the conceptualization as well as the resulting landscape.) The architectonic designs smack of hubris, and while skilled as formal manipulation, beg the question of ecological reason: plants fill out a shape or occupy a pattern, and mono-species set out in great number seems to be tempting fate. The naturalistic approach is hardly less problematic. We inhabit "landscapes," defined by J. B. Jackson as the product of human dwelling interacting with the land. That land is no longer "natural"; innocence ended with habitation. Thus, the problem becomes determining what nature is and what it directs. As the Dutch landscape historian Erik de Jong has stated, "there is no such thing as nature in the singular"—it is not that simple; there are many natures—"our conception of it is dependent on the historical and social context."[7] Nor is there any simple or single application of factors perceived as being natural.

Given that the projects submitted to the Designed Landscape Forum were experienced by the panel only as projected images and explained with limited accompanying written explanations, it is hardly any surprise that evaluation was based primarily on the designs' formal aspects. Of course, this is problematic, reducing landscape architecture to a visual phenomenon, discarding any appeal to the other human senses or to use. Nor were the slide submissions, for the most part, coherent or comprehensive. Few provided any sense of the immediate or greater context, and how the design fit as a piece of a greater entity. The designer's claims notwithstanding, it was virtually impossible, for example, to accept the ecological performance of certain land reclamations, or to dismiss more formally-conceived landscapes as being antisocial or inappropriate to their sites.

How much can the "look" of the landscape tell us about its validity and performance? Must an ecologically conceived design look "natural"? If so, why so? If the program calls for human recreation, the landscape is no longer functioning as it was in its "natural" state. What are viable forms for complex programs that interweave considerations of ecology, human activity, form and space, and horticulture? Are there strategies that appear more promising in generating landscape architecture for our times?

Within the intersections of these stray and wandering paths loomed another large question: where does the human occupant enter into these designs? Remarkably few of the presentations included people, leaving little clue to the intended perception or use. The sole exceptions to this dearth of human presence seemed to issue from the more corporate landscape design offices (usually sets of initials rather than the names of people), whose images smacked of the developer's promotional brochure in which everyone plays or shops happily together. We should not forget that even the most formal of garden and landscape designs constitutes a setting for people (and animals). Spiro Kostof, in his monumental history of world architecture, reminded us that architecture is inextricably bound in both "settings" and "rituals" —both the place and the event.[8]

We need also consider the seemingly more mundane matters of maintenance and control. As early as the 1950s, Thomas Church and Garrett Eckbo raised the question of maintenance in garden design, acknowledging that many people today have neither the time, nor the means, nor the desire to spend time on extensive upkeep.[9] In Eckbo's work this consideration, paired with considerations of areas of more intense use, often led to more paving than planting. Most of the designs reviewed, however, appeared to avoid the issue of care, implying that the state of the landscape is eternal, or that intense gardening is a desirable practice (which, of course, it may be for certain projects).

But landscapes do grow (or die). Change is inherent in all living systems, and in turn, becomes a key ingredient in landscape—as opposed to architectural—design. Few schemes addressed either growth or change; nor was there much to suggest the power of time to inform the history of the landscape. Why this avoidance of evolution? Why the eternal moment or acceptance of only four seasons? Must we always choose between process-oriented landscape design (which never seems to appear as tidy artifact) and the tidy design (which never seems to address process)? The merging of the two directions suggests fertile terrain for making landscapes in coming years.

For the immediate future, we could postulate at least two plausible directions. The first continues the notion of a formally ordered landscape, perhaps drawing on architectural or sculptural ideas, perhaps more accepting of environmental management practice. Rather than a reliance on mono-species planting that courts ecological disaster (when one tree dies, there is a great danger they will all die), or a design that demands the continued use of the spade and the shears, however, let us suggest a scheme in which the vegetation is allowed to take its course as best it can. Here, more in the Japanese manner of mixed formalities, the intersection of two orders—human and horticultural—can create a landscape of more than one dimension [12-1].[10]

The second approach looks more deeply into the ecological processes, and sees within them the suggestion of form that geometry or an artificial construct borrowed from the art world would never have invented. Beyond merely addressing the logic of drainage, erosion, growth, sunlight and orientation, prevailing breezes, and horticultural suitability, one may find in them—in combination—the generators of landscape form. The resulting aesthetic derives not simply from analysis, but analysis and understanding translated to form. If, in the past, we have used the shape, color, and texture of a single plant as the basis of planting design (think of the monochromy of the White Garden at Sissinghurst or, more generally, the planting of rose gardens), we would now look at the systemic workings of these ecological systems as instigators of the perceived world—used by designers not as a free-for-all, but as a landscape aesthetically and socially conceived. Early applications of this approach already exist. Gilles Clément's *jardin en mouvement* ("garden in movement") in one area of the Parc André Citroën in Paris sowed the seeds of grasses and wildflowers without any exact preconception of the form they would take over time. In an accelerated Darwinian evolution, the microclimatic and localized forces coerce an initial and constantly changing garden within the formal structure of the park as a whole. It is a daring experiment in which the designer, like an athletic coach, must be content to establish the parameters and then let the participants take charge.[11]

Waterfront park developments by Hargreaves Associates in Louisville and Lisbon harness the coercive force of water flow to derive planning at the macro-scale and more particular, localized landforms in accord with social activity.[12] In these cases, the designers have abrogated the idea of simply creating a naturalistic scene; instead, we enter a zone of managed hyper-reality, in which all the players appear familiar and yet the forms they take are novel. As more than one participant in the symposium stated, a landscape doesn't have to look "natural" to be E.C. (ecologically correct).

Why try to recreate a nature that is today nostalgic and out of step with current patterns of use? (We can't recreate Yosemite.) Can we instead imagine landscapes—avoiding replication—that will develop from a deeper understanding of natural systems and the human as one element—albeit a very powerful one—of those systems? This idea seems particularly relevant to sites despoiled by decades or centuries of industrial use. New uses, new trials, and new factors will determine the look of fabricated natures in the future. There is every indication that they will be designed and managed, whether as acts of preservation, creation, or interpretation.

Originally published in Gina Crandell and Heidi Landecker, eds., *Designed Landscape Forum 1*, Washington, D.C.: Spacemaker Press, 1998.

Notes

1 The Designed Landscape Forum was held in San Francisco in November 1995, organized by a concerned group of landscape designers, artists, and architects. This paper was presented on the panel concerning criticism.

2 There is a danger in this shift, however, as sociologist Robert Gutman has pointed out. Traditionally, a professional serves as an instrument for both the individual client and the public good. As the image of the design professional becomes more closely associated only with the desires of the client, the public's respect for the profession wanes. See Robert Gutman, *Architectural Practice: A Critical View*, New York: Princeton Architectural Press, 1988.

3 Ian McHarg, *Design with Nature*, Garden City, NY: Doubleday, 1969.

4 I have discussed this trend at greater length in "Form, Reform (and American Landscape Architecture)," in *Het Landschap/ The Landscape*, Antwerp: deSingel, 1995, pp. 37–52.

5 See Michael Benedikt, *For an Architecture of Reality*, New York: Lumen Books, 1987, and Marc Treib, "An Island in the Riptide toward Dissolution," *Journal of Architectural Education*, Volume 40, Number 2, Jubilee, 1987, pp. 78–79.

6 See Marc Treib, "The Place of Pattern," *Pages Passages*, Number 4, 1992, pp. 128–134.

7 Erik de Jong, "Nature in Demand: On the Importance of Ecology and Landscape Architecture," *Archis*, October, 1996, p. 65.

8 Spiro Kostof, *A History of Architecture: Settings and Rituals*, New York: Oxford University Press, 1985.

9 "There is of course, no such thing as a 100% maintenance-free garden, and if there were, you would soon be tired of it, for it would cease to be a garden. Most people mean they want the space so organized that they may know the delights of gardening in what little time they have; that they will not become slaves to a scheme that never looks at its best no matter how much time they labor on it." Thomas Church, *Gardens Are for People*, New York: Reinhold Publishing, 1955, p. 21. See also Garrett Eckbo, *The Art of Home Landscaping*, New York: McGraw-Hill, 1956, especially pp. 238–245.

10 The mixture, or embeddings, of ordering systems is inherent to historical Japanese design. See Marc Treib, "Modes of Formality: The Distilled Complexity of Japanese Design," *Landscape Journal*, Spring, 1993.

11 Gilles Clément, *Le Jardin en mouvement: de la vallée au Parc André Citroën*, Paris: Sens & Tonka, 1994.

12 See *Hargreaves: Landscape Works*, *Process Architecture*, Volume 128, 1996.

Afterword

My interest in landscape architecture dates almost precisely to 1971. Until then, I rarely looked at what lay outside buildings. And being the product of a modernist architectural curriculum, I believed—whether warranted or not—that the lands of good design were Japan and the nations of Scandinavia. Upon completion of my undergraduate studies I traveled abroad for my first time and studied in Finland for a bit over a year under the auspices of a Fulbright Fellowship. Scandinavian architecture became one of my research interests thereafter. Then it was time for Japan.

Travels in Japan in 1971 took me from the southern tip of Kyushu almost to the top of Hokkaido, through seven months of nearly constant travel. While my purported subjects of interest were castles and folk architecture, I thought it prudent to also visit the monuments of Japanese architecture, its palaces and temples, and urban complexes. Kyoto absorbed much of my time. To my surprise, I found that rather than looking *at* Japan's temples I was looking *through* them—since they lacked facades in the Western sense of the word. Beyond the building, neatly framed within the opening, was a courtyard or a garden. Many of these gravel spaces, with their raked patterns and sparse plantings, or their play of architectural shapes against those more naturalistic, were vividly compelling. I was often transfixed. They seemed modern, or at least a model for modernism, and they were beautiful. Somewhere in my reading, however, or perhaps in a visit to one of the imperial gardens, I learned that the dry gardens frequently found in Zen precincts were but one of several types, and that an equally vibrant tradition existed in gardens based more squarely on a naturalistic model.

Somewhere in my reading on these so-called "stroll gardens" of the seventeenth century I caught a passing reference to their affinities with the "picturesque" English landscape tradition. "Hmm...," I thought, "perhaps I'd better read about those to establish a context for my thinking about Japanese gardens." And somewhere in my reading about English gardens, perhaps in an article by Nikolaus Pevsner, I caught a reference to the French formal gardens, against which the English had reacted, at least to some degree, or at least as some would prefer to believe. Those next readings led to the Italian gardens of the Renaissance, and by that time I was involved with landscape history. All of these topics, to varying degrees of knowledge or expertise, appear in these essays, and several examples appear more often than they should.

One additional clarification. My study of twentieth-century landscapes came more directly from my familiarity with modern architecture. I soon realized that virtually no one was working on landscape architecture from this period in a substantial way. My most fortunate meeting and long association with Dorothée Imbert cemented that interest. We were both pulled into the vacuum, so to speak. A disappointing 1988 symposium in New York prompted me to organize a parallel program in Berkeley the following year, an event that eventually saw light as *Modern Landscape Architecture: A Critical Review* in 1993. Several books, a number of chapters, and various articles have followed in the wake.

And things have evolved from there, with a continued thirst to see more and learn more, and perhaps even write more.

We all have our weaknesses.

For information, inspiration, and/or influence, thanks…

NICHOLAS ADAMS

SVEN-INGVAR ANDERSSON

THORBJÖRN ANDERSSON

TADAO ANDO

LODEWIJK BALJON

GASTON BEKKERS

CHARLES BIRNBAUM

ALEXANDRE CHEMETOFF

ALAN COLQUHOUN

JAMES CORNER

ERIK DE JONG

MARCO DE MICHELIS

GEORGES DESCOMBES

GARRETT ECKBO

KEITH EGGENER

SUSAN RADAMACHER FREY

JOHN FURLONG

DAVID GEBHARD

ADRIAAN GEUZE

CHRISTOPHE GIROT

PHILIP GOAD

GERT GROENING

PAUL GROTH

HIROSHI HARA

GEORGE HARGREAVES

DIANNE HARRIS

KENNETH HELPHAND

RON HERMAN

RICHARD HERTZ

Acknowledgments

In order of appearance, more or less:

Miss Josephine Curto taught my 7th (or was it 8th?) grade Honors English class at Miami Beach High School and tried her best to get me to write creatively. We resisted—or maybe it was just me—expository form with all that business of a clear and detailed outline and all that writing restricted to third person. In standardized examinations, my aptitude always tested higher in math and science than in English, but my grades never seemed to work out that way. Physics and structures were the low points in my architectural/college education, but that's another story. I never liked to write, although I did like to talk. Maybe I liked talking too much (more than one person has suggested it), but that too is another story. Maybe it was just that writing and talking were so different, and I suppose that in some ways I have always been trying to get them into greater accord.

I remember reading an early essay by Tom Wolfe in which he explained how he found his voice by imagining writing a letter to his editor about the subject of his assignment. That made sense: there was an immediacy to a letter; there was a lack of pressure to get everything just right, even if simply stated. And a letter implied there would be a reader at the other end with an interest in what you wrote. In many ways the letter has been my model: a letter that began with an observation.

Before traveling to Japan I audited a course on Japanese gardens taught by landscape architect Ron Herman at the University of California, Berkeley, where I had recently joined the faculty. From him I learned a considerable amount about Japanese gardens and later, landscape architecture in general, including knowledge about the profession and plant materials ("Not nearly enough," he might say). Being roughly contemporary, we shared a similar educational background although based on opposite shores of the United States. Over the years we have had numerous hours of informative conversation, we taught a course on the architecture and gardens of Japan together at Berkeley and co-authored *A Guide to the Gardens of Kyoto*, originally published in 1980, and recently republished in a new edition. In some ways I need apologize to him for straying into his territory —but it is his fault, at least to some degree.

—

Martin Fox was the editor of *Print* magazine ("America's Graphic Design Magazine") for nearly four decades, and to Marty I really owe my start in writing for magazines on a regular basis. In the early 1970s, when I was producing a blizzard of no-budget, small-run posters for the Berkeley's departmental and college lecture series, *Print* selected three of them for their "Posters of the Decade" issue. An article on my work appeared subsequently as did a meeting with the editor in New York. We discussed various writing possibilities, at which time I told him of my interest in the vernacular commercial environment as well as more polite design practice. Thus began a stint as author and contributing editor for *Print* that lasted almost 15 years, and during which I vented my spleen on subjects such as postmodern graphics, or sang paeans to the Goodyear blimp's electronic signs, or in a three-part article tried to explain the form and semantics of signing in the urban environment. I am eternally grateful to Marty for providing that forum for linking architecture and graphic design, and for looking at the semantic aspects of the built environment.

Susan Radamacher Frey, as editor of *Landscape Architecture* magazine in the mid-1980s, accepted a series of articles that really brought me into the landscape design arena. Susan was open to almost any of my proposals, and she sponsored texts on plazas in Japan and modern cemeteries in Sweden, as well as another series on the vocabulary— I got as far as water, rock, and topiary—of landscape architecture.

When John Dixon Hunt served as director of the landscape studies area of Dumbarton Oaks, he invited me to serve there as a Senior Fellow. In my six years on the board I encountered a group of people concerned with more scholarly thinking about landscape architecture, scholars based in a very expansive range of disciplines, who helped broaden my scope of investigation. As editor of the *Journal of Garden History*, later *Studies in the History of Gardens and Designed Landscapes*, John published a number of my essays—two of which are reprinted here—and several overly long book reviews. This despite radically different backgrounds and even more radically differing tastes in landscape design. For all of the above, I am most grateful.

I should also note in passing that it was the late Elisabeth MacDougall —John's predecessor at Dumbarton Oaks—who accepted my first paper at a Society of Architectural Historians annual meeting, launching me, so to speak, on another facet of my academic career. That career, however, began with Gerald McCue, former chair of architecture

RANDOLPH T. HESTER, JR.

THOMAS HINES

WALTER HOOD

CATHERINE HOWETT

JOHN DIXON HUNT

DOROTHÉE IMBERT

ROBERT IRWIN

JOHN BRINCKERHOFF JACKSON

CARLOS JIMENEZ

TIBOR KALMAN

ERIKA KIENAST

SPIRO KOSTOF

MARK LAIRD

LUIGI LATINI

MICHAEL LAURIE

LARS LERUP

MARK LINTELL

WAVERLY LOWELL

ANNEMARIE LUND

KAREN MADSEN

FUMIHIKO MAKI

TOM MARTINSON

GUILHERME MAZZO DOURADO

MARY MCLEOD

ROBIN MIDDLETON

SHIRO NAKANE

LANCE NECKAR

LAURIE OLIN

THERESE O'MALLEY

JUHANI PALLASMAA

ALESSANDRA PONTE

ALAN POWERS

REUBEN RAINEY

PETER REED

ROBERT RILEY

STANLEY
SAITOWITZ

YOHJI SASAKI

GÖRAN SCHILDT

MARIO SCHJETNAN

MARTHA
SCHWARTZ

SUSAN
SCHWARTZENBERG

GEORGE SEDDON

HUGO SEGAWA

LELAND SHAW

TOM SIMONS

CHIP SULLIVAN

MAKOTO SUZUKI

SIMON SWAFFIELD

GEORGES TEYSSOT

JAMES TURRELL

KIRK VARNEDOE

PETER WALKER

JOACHIM
WOLSCHKE-
BULHMAHN

KEN WORPOLE

JAN WOUDSTRA

FEDERICA ZANCO

ELYN ZIMMERMAN

JOHN ZURIER

NINA HUBBS
ZURIER

at Berkeley and later Dean of the Graduate School of Design at Harvard University. He hired me—against my better judgement—to teach at Berkeley, where I have remained ever since. Stranger things have happened, I suppose.

Then there were those who helped shape my writing style as well as the ideas and the standards. Mary McLeod at Columbia University was as sharp a critic as one might (or might not) wish for, and in our younger and less busy days we exchanged manuscripts for mutual critical review. Mary was brilliant at perceptively critiquing subjects about which she knew virtually nothing and in spotting writing tics that eluded even the most diligent editor. She never forgave me for beginning articles with overly broad and sweeping declarations or for using the word "stuff," but we remain good friends nonetheless.

Le Corbusier once described himself as a lion who had eaten the flesh of so many different animals—so to speak—that it didn't really make sense to try to establish exactly which ones had had the greatest effect on his being. But there are three people to whom I owe an enormous debt and who can be identified. If I have ever truly had a mentor, it was my dear friend Spiro Kostof, who died in 1991. As an architectural and urban historian, Spiro possessed an enormous breadth of vision and depth of understanding. His conversations were stimu- lating, humorous, and always challenging, and his critiques of my written pieces were merciless (as well they should have been). He deigned to provide this service for me out of friendship, even if I was a "designer" (which a number of historians in the architecture field often use as a euphemism for a kind of mental retardation). Although born in Turkey of Greek lineage, and coming to the United States only at the postgraduate level, Spiro possessed a facility with the English language that remains unsurpassed by anyone I have encountered in architecture. His vibrant writings and dynamic lecture style were models for all of us (slightly) younger colleagues. He taught me many great lessons: don't describe without telling the reader why; don't tell a skeletal tale, but flesh it out with varying perspectives; situate your investigation in an ever-greater context. He qualified one writer we discussed over lunch one day as having a "subtle mind"; a quality and goal that sticks with me as I write. That notion of subtlety has encouraged me to look in a more multifaceted manner rather than restricting the argument to the one-dimensional and the obvious.

If Spiro represents one of my "pillars," John Brinckerhoff Jackson, who passed away in 1996, was the other. Over a period of nearly twenty-five years I continually learned from, and genuinely enjoyed, the great mind and the scholar. Although normally qualified as a "historian of the cultural landscape," Brinck had an incredible understanding of both the High and Low manners, an interest we shared although in differing ways. He saw all points of view and constructed stories in complex ways that might draw equally upon learned legal tracts and vernacular folk tales. He did not really write books: he was an "essayist" in the true sense of the word. There were neither real footnotes nor endnotes; he wrote in a sympathetic tone, at times with a wry or ironic humor. Most of all, he began with observations, using travel and seeing as a primary research method. In some ways, although in a manner quite distinct from my own, he validated my belief that study could begin with attending and emotion, rather than only by formulating a new theory or method. Or even better, address them all.

And most of all, I need acknowledge the contributions of Dorothée Imbert, with whom I shared over ten years of contact in all areas of our lives. During that time, at home and on the road, she provided thoughts, research, criticism, editing, joy, and pleasant and informative companionship on visits to numberless gardens and buildings. As a critic of my writing (and design work) she was fearless, despite the fact that French and then Italian—but certainly not English—were her native languages. Readers of most of the essays contained herein can be most thankful for her rigorous application of what we came to term politely as the "bullshit filter." I would find comments in the margins which read something like: "This sounds great: Do you have any idea what it really means?" The gauntlet was cast down and the challenge made; I must clarify or discard. Dorothée may think that over the days and years she got more from me than I did from her, but I'm pretty sure it was the other way around. Or at best, we could call it a draw (but definitely with her giving more than she got).

Thorbjörn Andersson and I have shared a passion for visiting landscapes, and the conversations on our many excursions have provided or stimulated ideas certain to have appeared in the writings contained in this book. A number of them are probably his although I have unintentionally passed them off as my own. His indulgence is begged. Georges Descombes has played a similar role, but since Georges gets the dedication for this book—the dedication tells it all—no more will be said here other than mentioning my enormous respect for his thinking and his built work.

Several friends agreed to review the rough selection of the writings to establish the final cut and shape of the collection. For their help in this department I thank Nicholas Adams, Dianne Harris, and Richard Hertz.

To Caroline Mallinder at Routledge I offer profuse gratitude, that is, now that the main work is done. It was through her instigation, encouragement (read: unbridled flattery), and good cheer that I undertook and saw through what has turned out to be a rather complex and entirely intimidating project. Thanks also to Katherine Morton and Susan Dunsmore at Routledge for helping the book see the light of day. And to Karen Madsen thanks for careful proofreading and last-minute editorial suggestions.

To all those many other people who have contributed in some way large or small, I can only say thank you, and try to remember their names as best I can. And I ask for forgiveness for those forgotten through no conscious intention of mine.

Illustration Credits

Index